Mainstreaming Solar Energy in Small, Tropical Islands

This book explores how cultural considerations can improve policymaking to achieve mainstream solar energy in small, tropical islands.

Focusing on Trinidad, Barbados and Oʻahu, Kiron C. Neale looks at how culture can affect and be affected by the policies that support the household adoption of two key energy technologies: solar water heating and photovoltaics. Drawing on interviews with residents and energy officials, and an examination of the institutional, socio-economic and physical factors that affect energy systems such as governance structures and energy resource availability, the author explores themes including the impact of insularity on energy transitions and behavioural and cultural change. Overall, this book rebrands policies as instruments of cultural change and puts forward recommendations applicable to all small, tropical islands.

Following the islands' transition to renewable energy, this book will be of great interest to scholars of energy policy, energy transitions, climate change, cultural studies and small states development, as well as industry professionals working on energy policy implementation.

Kiron C. Neale completed his postgraduate studies at Oxford University's Environmental Change Institute in 2018, and is currently involved in coordinating energy efficiency projects for small to medium-sized enterprises in Oxfordshire as the EiE Environmental Officer at Oxford Brookes University, UK. He is from Trinidad and Tobago in the Caribbean and grew up in Southern Trinidad (San Fernando and Princes Town).

Routledge Studies in Energy Policy

Low Carbon Politics
A cultural approach focusing on low carbon electricity
David Toke

Governing Shale Gas
Development, Citizen Participation and Decision Making in the US, Canada, Australia and Europe
John Whitton, Matthew Cotton, Ioan M. Charnley-Parry and Kathy Brasier

Business Battles in the US Energy Sector
Lessons for a Clean Energy Transition
Christian Downie

Guanxi and Local Green Development in China
The Role of Entrepreneurs and Local Leaders
Chunhong Sheng

Energy Policies and Climate Change in China
Actors, Implementation and Future Prospects
Han Lin

Energy Efficiency in Developing Countries
Policies and Programmes
Suzana Tavares da Silva and Gabriela Prata Dias

Ethics in Danish Energy Policy
Edited by *Finn Arler, Mogens Rüdiger, Karl Sperling, Kristian Høyer Toft and Bo Poulsen*

Mainstreaming Solar Energy in Small, Tropical Islands
Cultural and Policy Implications
Kiron C. Neale

For further details please visit the series page on the Routledge website:
www.routledge.com/books/series/RSIEP/

Mainstreaming Solar Energy in Small, Tropical Islands
Cultural and Policy Implications

Kiron C. Neale

First published 2020
by Routledge
2 Park Square, Milton Park, Abingdon, Oxon OX14 4RN

and by Routledge
52 Vanderbilt Avenue, New York, NY 10017

Routledge is an imprint of the Taylor & Francis Group, an informa business

© 2020 Kiron C. Neale

The right of Kiron C. Neale to be identified as author of this work has been asserted by him in accordance with sections 77 and 78 of the Copyright, Designs and Patents Act 1988.

All rights reserved. No part of this book may be reprinted or reproduced or utilized in any form or by any electronic, mechanical, or other means, now known or hereafter invented, including photocopying and recording, or in any information storage or retrieval system, without permission in writing from the publishers.

Trademark notice: Product or corporate names may be trademarks or registered trademarks, and are used only for identification and explanation without intent to infringe.

British Library Cataloguing-in-Publication Data
A catalogue record for this book is available from the British Library

Library of Congress Cataloging-in-Publication Data
A catalog record has been requested for this book

ISBN: 978-0-367-36456-4 (hbk)
ISBN: 978-0-429-34624-8 (ebk)

Typeset in Bembo
by Wearset Ltd, Boldon, Tyne and Wear

Printed and bound in Great Britain by
TJ International Ltd, Padstow, Cornwall

Dedicated to my renewable source of faith, the Lord my God, who made the Greater Light, the Sun.

Contents

List of illustrations	ix
Preface	xi
Acknowledgements	xiv
List of abbreviations	xv

PART I
Small islands, energy transitions and 'mainstream culture' — 1

1	'Sun, sea and sand', and solar energy	3
2	Energy transitions and the mainstream	28
3	Energy transitions and mainstream energy cultures	49

PART II
Beginning the household solar energy transition — 67

4	Agriculture, fossil fuels and electricity	69
5	Electricity and mainstream energy cultures	94

PART III
Transitioning to and through household solar energy technologies — 123

6	Electricity, solar hot water and mainstream cultural change	125
7	Solar water heating, PV and policy implementation	153

PART IV
Mainstreaming solar energy in small, tropical islands   185

8 Conclusions on mainstreaming solar energy   187

    Index   203

Illustrations

Figures

1.1	Showing the classic imagery of small, tropical islands	3
1.2	Showing the locations of the book's case studies	9
1.3	Showing a graphic summary of the book's energy transitions narrative	16
2.1	Showing the phases of an energy transition in the MLP	31
2.2	Showing the MLP and the conceptual positioning of mainstreaming	38
2.3	Showing the positioning of where the case study islands would fit in the MLP relative to their energy transitions through residential solar energy	39
3.1	Showing an example of the conventional expressions of 'culture'	50
3.2	Showing a conceptualization of the ECF	54
3.3	Showing three arbitrary energy subcultures embedded in the wider dominant energy culture	57
3.4	Showing 'culture' transcending the three planes of the MLP	59
4.1	Showing the export-oriented nature of Trinidad's energy system	78
4.2	Showing the relative significance of the characteristics policy officials believe are important for mainstreaming an energy innovation in Trinidad	83
5.1	Showing the household energy inputs which emerged from the inductive thematic content analysis of the interviews conducted with residents in Trinidad	96
5.2	Showing the perceived peak energy demand times of the residents interviewed in Trinidad	101
6.1	Showing the relative significance of the motivations for adopting solar water heating based on the inductive thematic content analysis of the interviews conducted with residents in Barbados	131

6.2 Showing the household energy inputs which emerged from the inductive thematic content analysis of the interviews conducted with residents in Barbados 135
6.3 Showing the perceived peak energy demand times of the residents interviewed in Barbados 137
6.4 Showing water being heated on a stove by combusting LPG 140
7.1 Showing the import-oriented nature of Oʻahu's energy system 158
8.1 Showing seagulls gliding in the sky along a tropical beach 202

Tables

4.1 Showing the deductive coding structure related to the barriers to solar energy implementation in Trinidad 82
4.2 Showing the extent of Trinidad's residential solar energy policy development 84
5.1 Showing the electricity category's coding structure which forms part of the 'energy supply' cultural domain 97
5.2 Showing the water heating material culture of the residents interviewed in Trinidad 104
6.1 Showing the retail market's coding structure related to the types of solar hot water information that the residents interviewed in Barbados wanted before adopting their SWHs 127
6.2 Showing the solar energy category's coding structure which forms part of the 'energy supply' cultural domain 136
7.1 Showing Oʻahu's power generation capacity 156
8.1 Showing key energy cultures policy goals that can be associated with a FiT 195

Preface

'Sun, sea and sand' epitomizes the classic imagery of small, tropical islands. However, such islands are highly vulnerable to the implications of climate change and energy insecurity. The 'sun' in the above imagery is one of these islands' potential sources of energy and in that light (pun intended), using the sun's energy, a renewable resource, can mitigate the greenhouse gas emissions linked to climate change, improve energy independence and economic self-sufficiency and develop more resilient societies in small, tropical islands.

My home, Trinidad and Tobago, is uncharacteristically well-endowed with fossil fuels (by island standards) and has no significant solar energy usage. The fossil fuel industry has several links to the islands' culture, for example the national instrument, the steelpan, is traditionally made from old oil drums. This sort of seemingly subtle hydrocarbon-culture connection was the initial inspiration for my interest in 'culture' and 'energy' – and this fed into my pre-existing interest in 'island solar' which I had since my undergraduate studies at the University of the West Indies. Further, the policymaking capacity and policy implementation in small, tropical islands are often constrained and not seamless (but then again, where isn't?).

I began my doctoral studies in 2014 so *Mainstreaming Solar Energy in Small, Tropical Islands: Cultural & Policy Implications* is a book that has taken almost six years to come to fruition. It looks at how 'culture' can affect and be affected by the policies that support the adoption of household solar energy in small, tropical islands. It is aimed at demystifying the meaning of 'culture' to policymakers and rebranding policies as instruments of cultural change.

Two of the most popular and established solar energy technologies globally are solar water heaters (SWH) and photovoltaics (PV) – the first uses the sun's energy to heat water and the second uses it to generate electricity. Mainstreaming household SWHs and PV in small, tropical islands is rooted in the idea of an 'energy transition' occurring in the residential sector due to the mass adoption of the technologies.

An energy transition here means a fundamental systems-level change that affects all the dimensions that make up the energy supply-demand space, that is, industries, markets, cultures, technologies, science and policies – what Frank Geels' work conceptualizes as 'the socio-technical regime' in the multi-level perspective (MLP) of socio-technical energy transitions, the theory used to approach energy transitions.

The book defines energy transitions, the socio-technical regime and mainstreaming as theoretical starting points for understanding how 'culture' and policies interact. In addition, the role(s) of 'culture' in energy transitions and the MLP was not well-articulated or defined in the literature. Consequently, Janet Stephenson's Energy Cultures Framework was used to define 'culture' in the residential electricity regime as an 'energy culture' made up of norms, practices and material culture as well as the external influences of these features.

Best practices, lessons learnt and knowledge transfers are oftentimes adapted from the Global North or continental geographies in islands, that is, a vertical perspective looking northwards from the South. But looking laterally at other islands may hold more potential for meaningful exchanges. So, the narrative uses the islands of Trinidad, Barbados and Oʻahu as case studies to look at the previous theoretical insights.

These select islands are at three different points in their residential (solar) energy transitions and represent the contemporary extent of such transitions: Trinidad has no significant household solar energy adoption; Barbados has mainstreamed SWHs and is transitioning to/through distributed PV; and Oʻahu has mainstreamed distributed PV and is transitioning to/through battery-integrated PV.

The book is based on my doctoral thesis and the research used structured interviews with residents in Trinidad and Barbados, and semi-structured interviews with policy officials in Trinidad and Oʻahu. The residential interviews focused on the cultural formations related to energy and the institutional interviews focused on themes related to solar energy policymaking. The interviews were respectively analysed using an inductive and deductive thematic content analysis.

Finding an apt narrative using the data collected for three different islands where only cultural and policy-based data was collected for one, Trinidad, and only cultural data for Barbados and only policy-based data for Oʻahu has been perhaps the key challenge in terms of making the lessons as clear and relevant for all small, tropical islands. But a historical energy transition chronology is the best way to convey the story – a systems-level story of complexity, transience and evolution. Therefore, the narration follows the islands' solar technological timeline, that is, a fossil fuel-powered residential electricity sector transitioning to include SWHs followed by PV (from having used resources such as biomass and metabolic energy before fossil fuels pervaded).

The book improves the spatial understanding of the MLP by applying it to not one but three case studies. Beyond this, by being applied to small, tropical islands, it represents a highly specific geographic application that is rather novel. Small, tropical islands are also amongst the most vulnerable geographies relative to the repercussions of energy insecurity and climate change. So, this islandic research on residential solar energy is a valuable contributor to understanding the mainstreaming of alternative energies in these marginal geographies.

Additionally, the book's exploration of the roles of 'culture' in energy transitions which is coupled with the academic application of mainstreaming as a concept to consider in the MLP shed light on how residential solar energy policies can be humanized – which will prove useful to governments keen on implementing initiatives which support residential solar energy.

In the wider scope, the book is useful reading for persons working in, or interested in themes related but not limited to climate change mitigation, energy security, the environment, energy and renewable energy, energy transitions and sustainability. The content is also useful reference material for students of geography, environmental management and energy-related disciplines.

Kiron C. Neale
Oxford, United Kingdom

Acknowledgements

I wish to first offer thanks to God, the source of my faith; to my mom, Fermanda Mohammed, and dad, Clive Neale (may he rest in peace), for their love and sacrifice over the years and for always believing in my dreams and ambitions; to Carmen Alonso for her continued love and support; and to my friends and family for shaping me.

Thanks especially goes out to my doctoral supervisors, Professor Nicholas Eyre and Dr Christian Jardine, for their time, energy and guidance throughout the development of the research on which this book is based – I am so grateful to have been under their tutelage. In that same breath, thank you to my doctoral examiners, Dr Rebecca Ford and Professor Benjamin Sovacool, as well as all those others too numerous to mention who provided wider academic support through critiques, brainstorming and moral encouragement.

I extend my heartfelt gratitude to all the householders and officials (as well as their organizations) and contacts in Trinidad, Barbados and Oʻahu who were a part of the interviews which provided the primary data that informed the doctoral research and book's narrative. Also included here are those responsible for the various secondary data sources and literature which provide a working foundation for researchers such as myself.

I wish to thank Routledge and, particularly so, Annabelle Harris, the Editor, as well as Matthew Shobbrook, the Editorial Assistant, of Routledge's Environment and Sustainability list for their engagement, guidance and, simply, for giving me the opportunity to publish my work. These thanks also extend to the three anonymous reviewers who provided valuable commentary, critiques and feedback on the book proposal.

A special note of thanks finally goes to The Rhodes Trust, the University of Oxford and Santander Bank for providing the funding which made the various doctoral research projects in the three islands possible.

Abbreviations

AC	alternating current
BL&P	Barbados Light and Power Company
BNOC	Barbados National Oil Company
BOE	barrels of oil equivalent
BREA	Barbados Renewable Energy Association
CGS	Customer Grid-Supply
CO_2	carbon dioxide
CSS	Customer Self-Supply
DC	direct current
DCA	Division of Consumer Affairs
DPP	Department of Planning and Permitting
ECAC	Energy Cost Adjustment Clause
ECF	Energy Cultures Framework
EEPSs	energy efficiency portfolio standards
EID	Electrical Inspectorate Division
EPPD	Environmental Policy and Planning Division
FCA	Fuel Cost Adjustment
FiT	feed-in tariff
FTC	Fair Trading Commission
GDP	gross domestic product
GW	gigawatts
GWh	gigawatt-hours
HCEI	Hawaii Clean Energy Initiative
HECO	Hawaiian Electric Company
HEI	Hawaiian Electric Industries
HELCO	Hawaii Electric Light Company
HNEI	Hawaii Natural Energy Institute
HSEA	Hawaii Solar Energy Association
HSEO	Hawaii State Energy Office
HSL	Hawaii State Legislature
IPP(s)	independent power producer(s)
km	kilometres
km^2	square kilometres

kW	kilowatt
kWh	kilowatt-hours
kWh/m^2	kilowatt-hours per square metre
kWp	kilowatt-peak
LED	light-emitting diode
LNG	liquefied natural gas
LPG	liquefied petroleum gas
m^2	square metres
MEAU	Multi-lateral Environmental Agreements Unit
MECO	Maui Electric Company
MEEI	Ministry of Energy and Energy Industries
MLP	Multi-level Perspective of Socio-technical Energy Transitions
mm	millimetre
MPD	Ministry of Planning and Development
MPU	Ministry of Public Utilities
MW	megawatts
NEM	Net Energy Metering
NGC	National Gas Company of Trinidad and Tobago
NPC	National Petroleum Corporation
°C	degrees celsius
OPEC	Organization of the Petroleum Exporting Countries
PowerGen	Power Generation Company of Trinidad and Tobago
PPA(s)	power purchase agreement(s)
PPLS	Petroleum Production Levy and Subsidy Act
PUC	Public Utilities Commission
PV	photovoltaics
REEEA	Renewable Energy and Energy Efficiency Agency
RER	Renewable Energy Rider
RETs	renewable energy targets
RIC	Regulated Industries Commission
RPSs	renewable portfolio standards
SIDS(s)	Small Island Developing State(s)
SPT	Social Practice Theory
SWH(s)	solar water heater(s)
T&T	Trinidad and Tobago
T&TEC	Trinidad and Tobago Electricity Commission
TGU	Trinidad Generation Unlimited
U.S.	United States of America
UofH	University of Hawaii
UTT	University of Trinidad and Tobago
UWI	University of the West Indies
VAT	value added tax
W	watt(s)
WW(s)	World War(s)

Part I
Small islands, energy transitions and 'mainstream culture'

1 'Sun, sea and sand', and solar energy

Figure 1.1 Showing the classic imagery of small, tropical islands.

Imagine yourself standing at the edge of the treeline in Figure 1.1 looking at the beach whilst the sunlight ever so lightly strikes your face through the gaps in the leaves of the coconut trees overhead.

Now, move further down onto the beach and sit for a bit. Imagine that you're now sitting on the warm, white sand and just in front of you, the gentle waves lap the sands ever so lazily.

Cast your gaze from your feet towards the horizon and take in the way in which the nearby clear water blends into the aquamarine waters of the coral shallows then fades into deeper hues of blue beyond the surf.

4 *Island energy transitions and cultures*

Soon after, a tropical breeze blows and rustles the leaves of the trees behind you and tingles your skin. Suddenly, you start noticing the tan lines on your sun-kissed arms and legs; it's a gentle reminder of how much sunshine the island gets so you think to yourself:

'*The sunshine is so warm … so radiant … so, quite frankly, "tropical"!*'

The nature of small, tropical islands

The above is the type of touristic imagery that has popularized 'sun, sea and sand' as a branding for small, tropical islands. But the sunlight in the above scene is an energy resource. So, what if these islands were to mainstream this energy? To begin answering this question, it is worth characterizing small, tropical islands.

Finding an absolute figure for the number of islands in the world is a hard, if not impossible task. But there are easily millions of them and perhaps only 11,000 that are permanently inhabited (Miaschi, 2017). There are numerous ways of defining islands and this would affect the numbers that are recognized. But for this book, oceanic islands are of interest and there are about 2,000 (Miaschi, 2017).

Whilst islands' geopolitics can affect the definition of 'small islands', their geographic features are quite similar. Therefore, the book uses these to drive the scope of the 'smallness' of islands. Kakazu (2007) states that defining islands' 'smallness' is relative and can refer to geography, demography, the economy or combinations of these.

But despite the multiplicity of definitions, the literature (e.g. Wong et al., 2005; Wong, 2011; Everest-Phillips, 2014; Baek, Kim & Chang, 2015; Ioannidis & Chalvatzis, 2017) states that most small islands have defining characteristics that include: limited natural resources; colonial legacies; mixed cultures; small populations; high population densities; rapid population growth; high emigration; high biodiversity; high dependencies on tourism, external aid and governmental interventions; diseconomies of scale; high energy and electricity grid connection costs; vulnerable energy infrastructures; high dependencies on fossil fuel imports; costly shipping; environmental fragility; oceanic natures; remote locations; and small land masses.

Of these, oceanic natures, remoteness and smallness intersect to create islands' insularity (Wong et al., 2005; Kakazu, 2007). However, Wong et al. caveat that the stronger an island's connectivity to the 'outside', the less distinctive its insularity. They further outline that these characteristics create an isola effect, that is, the physical isolation of islands such that they are exposed to different kinds of marine and climatic disturbances and have more limited access to space, products and services when compared to most continental locations.

Such conditions create smaller, less diversified yet more specialized economies because limited resources constrain production and consumption, and skew trade such that a wide range of goods/products are imported and only a few primary goods/products are exported (Wong et al., 2005; Kakazu, 2007; UN-OHRLLS, 2011; Wong, 2011).

For small, tropical islands specifically, these dynamics play out on islands located between the Tropics, that is, between the Tropic of Cancer and Capricorn in the Northern and Southern Hemispheres at roughly 23.5 degrees respectively. This is important because the Tropics are bathed in sunlight all year. Therefore, they have significant amounts of solar energy available to them.

But why is the availability of these solar energy resources important? The answer can be found by thinking about the small-island energy challenge – which can be best understood through two concepts: climate change and energy insecurity.

The small-island energy challenge

Small islands are marginal contributors to climate change yet are amongst the most vulnerable to its repercussions (Wong, 2011) – especially when compared to continental geographies (Szabó et al., 2015). But where climate change and small islands are concerned it is intuitive that sea level rise is perhaps the most famous implication. Others include potentially increased frequencies and intensities of climatic disasters such as hurricanes, loss of endemic species and bleaching of coral reefs.

Additionally, the world's energy is dominated by hydrocarbon markets and most small islands have little to no indigenous fossil fuel resources. Due to this scarcity they rely on costly hydrocarbon (usually oil/petroleum) imports (UN-OHRLLS, 2011; Dornan, 2015).

In the case of small, sovereign island states for example, they spend as much as US$90 million per day importing more than 900,000 barrels of oil (Rogers, Chmutina & Moseley, 2012, 2). These costs are passed on to local energy markets, so islands have higher than average costs when compared with non-islandic geographies as well (Wolf et al., 2016).

The IEA (2019) defines energy security as the uninterrupted availability of energy resources at affordable prices. So, energy insecurity in the context of small islands refers to a dependence on physical access to energy imports as well as high energy import bills and local energy costs – both of which are coupled to the fluctuating prices on international hydrocarbon markets. Therefore, given that hydrocarbon prices as well as supplies have been historically unpredictable, this puts the future of small islands' energy in a precarious position.

Though adapting to climate change has been prioritized by small islands, the implications of energy insecurity mean that renewable energy mitigation strategies are a potential solution to both issues (Verdolini & Galeotti, 2011) even though these islands' mitigation will be negligible globally (Dornan, 2015).

There is also an argument for saying that adopting renewable energy is also a form of adaptation since several technologies like household PV have been able to provide power to some homes when power outages occur because of the passage of a hurricane for instance. So, renewable energy will increase small islands' resilience.

Many small, tropical islands have several renewable energy resources that can be utilized. Using renewables reduces islands' carbon emissions, dependence on costly, imported hydrocarbons as well as energy costs in their local markets. This simultaneously improves their energy and economic independence as well as overall self-sustaining capacity since the growth in renewable energy is quite often linked to growth in indigenous energy use (Ioannidis & Chalvatzis, 2017).

The shift from fossil fuel energy systems to ones that incorporate more renewables represents an energy transition. In a nutshell, an energy transition is a shift in a system's combination of energy resources and technologies (Fouquet & Pearson, 2012) which structurally affect the technologies; localization of activities; consumption; science; education; manufacturing; and transport (Kemp, 1994; Cassen, Waisman & Hourcade, 2012).

Large-scale, centralized energy infrastructures are the dominant systems in most of the world and decentralized renewable energy systems exist in niches (Markard & Truffer, 2008). But the transition from centralized to decentralized systems is gaining prominence (Fuchs & Hinderer, 2016). The small scales of decentralized technologies are ideal for small islands so studying islandic energy transitions is a valuable contribution to energy futures research; islands are small-scale spatial laboratories (King, 2009) and testing sites for energy innovations (Ioannidis & Chalvatzis, 2017) but are often overlooked opportunities for energy research (Lilienthal, 2007; Shirley & Kammen, 2013).

So, where does solar energy fit into this idea of small, tropical islands and energy transitions? In order to answer this question, it is important to narrow down the elements that the book will focus on given the complexity of energy transitions.

Small-island (solar) energy transitions

'Culture', 'technology' and 'policy' as the key elements of interest

'Technology' is a key element of society and energy transitions, and changing technological systems changes who we are, our behaviour and our lifestyles (Miller, Iles & Jones, 2013). But an energy transition will not happen on its own (IEA and IRENA, 2017). Governments have multiple roles (Dirks et al., 2014) and they will need to support the niches for low-carbon energy sources and technologies (Fouquet, 2011) through the appropriate policy frameworks that can support the wider transition (IEA and IRENA, 2017). Therefore, energy policies will play an influential role in the technological changes that lead away from fossil fuel lock-ins (Patwardhan et al., 2012).

Arnesen (2013) argues that policies and regulations should include interventions related to socio-cultural energy consumption because they can act as signals that influence how social priorities are expressed through norms and habits. But there is a need to understand behavioural drivers to better design policy interventions (Egmond, Jonkers & Kok, 2006), so policies should quite simply be humanized (Hinchliffe, 1995).

Further, some have argued that the energy transitions work is largely mechanistic (Truffer & Coenen, 2012); there is an overemphasis on technical research in relation to policy formulation and implementation (Barron & Sinnott, 2013); and the academic discussion of transitions is largely based on techno-economic models (Sovacool & Geels, 2016).

These considerations suggest that there is a human element missing from technological research and policymaking. There are many human dimensions that could potentially apply but 'culture' is of interest. It has neither been concretely explored by the transitions nor wider energy literatures and has been rather implied (Stephenson et al., 2010; Strauss, Rupp & Love, 2013; Arnesen, 2013). Additionally, its role and influence(s) on (technological) energy innovations or energy policies are not well-researched relationships (Hinchliffe, 1995; Cherry et al., 2014).

This is important because the policies that influence innovation-adoption can accelerate or prolong energy transitions (O'Connor, 2010), and 'culture' is a barrier to but also enabler of implementation (Otte, 2014). Arnesen (2013) outlines that how energy is culturally interpreted should be a starting point for introducing innovations since generic policies being implemented in specific cultures often do not work. Likewise, innovations need to be adapted to local contexts (Geels, 2004) because introducing them into a social system where there is no grasp of its cultural background can have unexpected repercussions (Arnesen, 2013).

Based on these thoughts there are three central elements: 'culture', 'technology' and 'policy'. They do not exist in isolation (see Figure 2.1) but there are interactions between 'policy' and 'technology', 'culture' and 'technology', and 'culture' and 'policy'. The last are not as clear as the former two because the influence of 'culture' can be explored in the present, as well as an evolutionary element that enhances the temporal sensitivity of policies (see Foxon, Hammond & Pearson, 2010, 1206) since cultures change with time and in non-homogenous ways (Stephenson, 2018).

So, considering the above, the book's primary aim is to understand the role(s) of 'culture' in energy transitions by looking at the interactions between 'culture' and 'policy' during the mainstreaming of solar energy technologies. It seeks to shed light on how 'culture' influences and is influenced by energy transitions and to position solar energy policies as facilitators of cultural change.

The technological scope

Solar energy is one of the most widely implemented and viable renewable energy options globally to the extent that the market penetration pattern for solar systems is a useful indicator of the general transition towards using renewable energy resources (Sawyer, 1982). However, the book is specifically interested in household SWHs and PV – though the latter is the primary technology of interest given the dominance of electricity in modern energy systems. This is because of four key reasons.

First, SWHs and PV are relatively old renewable energy technologies; solar collection was around since the late 1700s, solar water heating from the late 1800s and PV from the 1950s (U.S. Department of Energy, 2004). They have undergone generational research and development to a point where today's systems are more efficient and cheaper (and will become increasingly so).

Second, solar energy technologies have been popularized throughout the world (Kabir et al., 2018) and SWHs and PV are arguably the most widely adopted and recognizable technological applications – as shown by their global growth trends. For example, between 2006 and 2016, the global PV capacity went from 6 GW to 303, and for SWHs, it went from 123 GW-thermal to 456 over the same period (REN21, 2017, 28 and 29).

Third, household SWHs and PV can potentially be more socio-culturally impactful. This is important because the book looks at individual households as the adopters. These technologies are decentralized, smaller-scaled energy technologies which means that compared to centralized, large-scaled technologies, there are more direct interactions occurring between an adopting household and their solar system; the gap between the energy source and use is shorter.

Fourth, household solar systems' small scales mean they can be installed on the roofs of homes. This reduces the opportunity costs associated with land use for large-scale solar energy projects in geographies that, by their very definition, have limited land space.

So, the 'mainstreaming' in the book's title refers to SWHs and PV being more widely adopted by households to the extent that they can be branded as mainstream technologies in the residential electricity systems of small, tropical islands. However, given the numerous small, tropical islands in the world, studying all these islands would be impossible.

Therefore, three case studies were selected which capture the contemporary extent of (residential) energy transitions and mainstreaming of household solar energy technologies: Trinidad, Barbados and Oʻahu (see Figure 1.2). As the figure implies, all three islands are within the Tropics, so they are well-endowed with solar energy.

Neill and Granborg (1986) suggest that before installing any renewable energy system the resource's potential should be considered. In that regard, the amount of incident solar radiation is one of the most important factors for using solar technologies and it is measured in kWh/m^2 per time (WEC, 2016).

Trinidad receives an estimated 2,102 kWh/m^2/year on a horizontal surface, Barbados receives 2,228, and Oʻahu gets 2,158; these figures were calculated using NASA's (2019) data. This shows that the islands have similar resource potentials. These figures can also be contextualized by considering that locations in Germany such as Freiburg which have significant amounts of

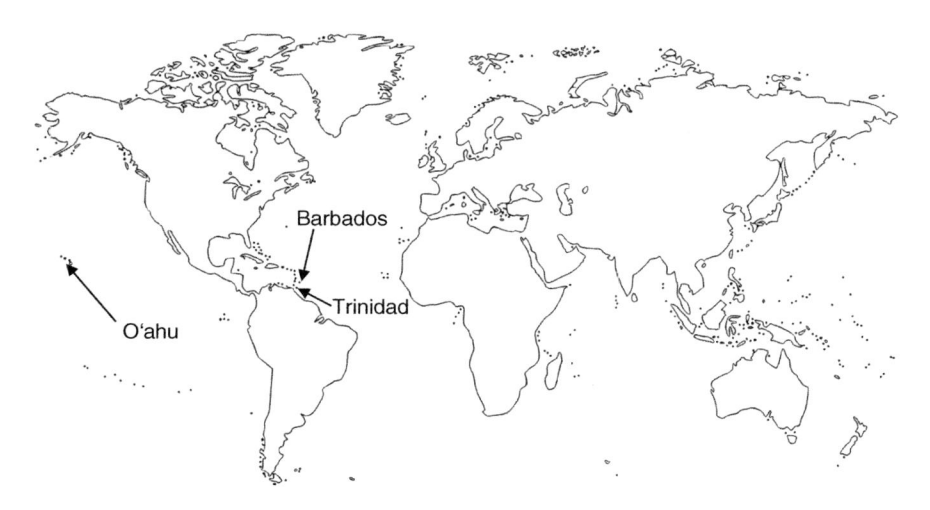

Figure 1.2 Showing the locations of the book's case studies.

household PV adoption (Dharshing, 2017), receive 1,117 kWh/m^2/year, for example (Thomas, 2012).

Despite small, tropical islands' abundant solar resources, if there is no access to the harnessing technologies, then the availability of the energy resource does not amount to much (Khan & Obaidullah, 2008). The issue is therefore being able to fully utilize the rich resource available (Chivers, 2015) and as will be shown, each case study has utilized their available solar energy resources to different extents. But in short, Trinidad has no significant amounts of household SWH or PV adoption; Barbados has mainstreamed SWHs and is transitioning to/through PV; and O'ahu has mainstreamed PV and is transitioning to/through batteries.

These thoughts are particularly useful because it shows that there is no resource disparity which may account for the case studies' varied extents of household solar energy adoption. So, what factors would account for this? Why are Trinidad, Barbados and O'ahu even being used as case studies?

A brief background on the case studies

Trinidad

T&T is one country but two islands. These are the Caribbean's southernmost islands, are located approximately 11 km off the coast of Venezuela, and cover approximately 5,100 km^2 (Espinasa & Humpert, 2016, 5).

T&T experiences a dry (January to May) and wet season (June to December), and the latter is markedly influenced by tropical depressions like hurricanes

(Wellington, 2011; Narinesingh et al., 2013). The islands have an average temperature of just under 28°C but it can range from 24 to 32°C (Narinesingh et al., 2013, 10). The annual rainfall also ranges from between 1,400 to 2,600 mm (Narinesingh et al., 2013, 8).

T&T is the Caribbean's leading producer of oil and gas and has been regarded as the most industrialized territory in the English-speaking Caribbean (Mustapha, 2012; Narinesingh et al., 2013; Solaun et al., 2015). The country's fossil fuel resources are such that peak oil production was recorded at 240,000 barrels per day and peak natural gas production was 746,700 BOE/day (Espinasa & Humpert, 2016, 7).

These resources account for between 39 to 45% of the GDP and 80 to 83% of its economy-wide exports (Mustapha, 2012, 29; Espinasa & Humpert, 2013, 57; Narinesingh et al., 2013, 59; Solaun et al., 2015, 3; GORTT, 2016, 2); the country has a US$22 billion-dollar economy (Statistics Section, 2019).

There are two main reasons why T&T an interesting case study energy-wise:

- *T&T is a significant greenhouse gas emitter by small, island standards*: Not only is T&T's CO_2 emissions an exception to the Caribbean trend of comparatively low emissions by global standards, but it is also a standout case for the CO_2 impact amongst SIDS (Thomas, 2009); its emissions per capita are so high that at one point it was ranked second in the world (Marzolf, Cañeque & Loy, 2015, 7).
- *T&T has low renewable energy adoption*: There is no literature that concretely documents the application and potential of renewables in Trinidad (Solaun et al., 2015). But, the MEEA (2012) and Khan and Khan (2017) state that T&T should invest in renewable energy because it can enhance energy security; mitigates climate change; uses reliable energy sources; creates employment; diversifies the economy; reduces the dependence on finite hydrocarbons; and promotes national and industrial development. Nevertheless, T&T has had no substantial renewable energy development to date; the Government is now beginning to consider how best to implement T&T's renewable energy assets and as is the case in many parts of the world, solar energy will require governmental support to be competitive (WEC, 2016).

However, Trinidad is the case study and not T&T. Trinidad is specifically being looked at because:

- *The nation's capital is in Trinidad*: Trinidad is home to the country's capital, Port of Spain, so it is the centre of policymaking and governance.
- *Trinidad is the more populated and urbanized*: T&T has 1.4 million residents (Statistics Section, 2019) and is more than 70% urbanized (Beaie, 2009, 22; MPSD, 2014, 9). However, over 1.2 million people live in Trinidad versus just under 61,000 in Tobago (CSO, 2012, 5). Further, T&T has

over 400,000 occupied, private housing units of which over 380,000 are in Trinidad (CSO, 2012, 33).
- *Trinidad has the larger residential electricity customer-base*: T&T's electric utility company, the T&TEC, has over 385,000 residential customers (Marzolf, Cañeque & Loy, 2015, 13; RIC, 2017, 6). Based on the earlier population and housing distribution, it is expected that Trinidad has the larger share of this customer-base; the RIC's (2016, 7) figures even show that Trinidad has roughly 94% of the T&TEC's total active customer accounts as well.
- *Trinidad is the more industrialized*: Tobago neither produces oil and gas nor is it as industrialized (e.g. oil-refining, ammonia production, etc.) as Trinidad.
- *Trinidad has the larger power generation capacity*: Trinidad's power generation capacity is rated at nearly 2,428.7 MW, and Tobago's is rated at 85.7 MW (Espinasa & Humpert, 2016, 8).
- *Trinidad has the higher energy demand*: Based on the preceding points, Trinidad has more residents, households, technical power generation capacity as well as residential electricity customers than Tobago. Therefore, it is expected that it has a higher residential electricity demand than Tobago.

Barbados

Barbados is a sovereign island-State in the Southern Caribbean. It is the Easternmost Caribbean island, is about 160 km from its nearest neighbours, and roughly 402 km north-east of T&T (Jensen, 2000, 93; MPDE, 2001; Ministry of Environment, Water Resources and Drainage (MEWRD), 2010, 8; Moore et al., 2014) – see Figure 1.1. Barbados also has an area of 431 km² (Moore et al., 2014, 15).

Barbados has an average temperature between 20 to 30°C (MEWRD, 2010, 8). Additionally, its annual rainfall varies from 2,000 mm in the wettest areas to 1,300 mm in the driest though most locations get between 1,400 to 1,600 mm (MEWRD, 2010, 8).

Like T&T, Barbados experiences a dry (December to May) and wet (June to November) season (MPDE, 2001; MEWRD, 2010). The climate is also affected by trade winds, tropical depressions, easterly waves and hurricanes (MPDE, 2001; Schlegelmilch, 2010) – which isn't surprising given that both T&T and Barbados are Caribbean neighbours hence their similar climates.

Similarly to Trinidad as well, Barbados is an oil and gas producer. But it only produces between 700 to 1,000 barrels of oil per day and between 290 to 500 BOE/day of natural gas which are shipped to Trinidad for refining and later imported (Gischler et al. 2009; Castalia Limited, 2010; SEforALL, 2012; Samuel, 2013; Ince, 2017, 2018); Barbados imports over 9,000 BOE/day (Gischler et al., 2009; Castalia Limited, 2010; SEforALL, 2012; GoB, 2015; Thompson, 2015, 6).

Barbados has a US$4.7 billion-dollar GDP economy (World Bank, 2019, 3); it is finance, insurance, business and tourism-oriented with a growing offshore financial services sector (MPDE, 2001; Schlegelmilch, 2010; MEWRD, 2010; Samuel, 2013). However, light industry and tourism make up the largest portion of the GDP and the latter is the most significant foreign exchange earner and economic driver (Schlegelmilch, 2010; Jackman & Lorde, 2012; GoB, 2013).

Barbados is being used as case study because:

- *Barbados is a geographical neighbour to Trinidad*: Barbados and Trinidad are both neighbouring Caribbean islands.
- *Barbados is a regional leader in residential solar energy*: Barbados is a pioneer in Caribbean sustainability (Haraksingh, 2001). It has been hailed as a regional leader in energy sustainability and a global leader in solar water heating because of its widespread adoption of SWHs and the establishment of an indigenous manufacturing and retail industry (Samuel, 2013; IEA-ETSAP and IRENA, 2015; Gray et al., 2015; GoB, 2015; Ince, 2017).
- The Barbadian SWH industry and supporting policies were around since the 1970s (Bugler, 2012; Rogers, Chmutina & Moseley, 2012; Samuel, 2013; CDB, 2014; Husbands, 2016). This means that there is an industry that can manufacture, retail, install as well as repair the technologies being put on the market in addition to there being the political support for it; the technology and its services were made readily accessible to the island's residents.
- As a result of such a retail market, access to the technology, available policies and subsidies, and international developments, 80 to 90% of Barbadian households have SWHs (Gray et al., 2015, 11; CEIS, 2015; IEA-ETSAP and IRENA, 2015, 1). Further, Barbados' market accounts for between 55 to 60% of the Caribbean's SWHs and 80% of its manufacturing (MPDE, 2001, 98; Bugler, 2012, 1). Barbados' solar water heating market is therefore saturated (IEA-ETSAP and IRENA, 2015).
- *Barbados is socio-culturally comparable to Trinidad*: Trinidad and Barbados were British colonies; were sugar-producing economies; are English-speaking with similar dialects; have strong African heritages due to the influence of slavery which include folklore and demographics; and share similar climates. They even have similar levels of economic development (Mustapha, 2012). It is therefore reasonable to assume that this common background keeps the islands' socio-cultural contexts and variables relatively comparable.
- The pervasive adoption of SWHs in Barbados makes it likely that solar water heating impacted the residents' culture that is associated with their household energy. So, given that the wider cultural context between Trinidad and Barbados should be more or less comparable, this leaves room to specifically look for differences between the cultures of residents in Trinidad and Barbados which may be attributable to the adoption of

solar water heating and which may influence the adoption of solar water heating. This is a useful approach since van de Vijver and Leung (1997) state that if identifying differences is the objective, then looking at similar cultures is a good starting point.

As the last point illustrates, Barbados is primarily being used as a cultural case study given its similar socio-cultural context to Trinidad's; it uses solar water heating as the case technology to investigate how 'culture' can potentially influence and be influenced by energy transitions – and in this case, an energy transition to/through SWHs. It is not being used as a policy-related case study because it does not have any significant adoption of household PV yet, and Oʻahu has a longer history of residential PV utilization as it relates to elements such as policies, markets and industries for example.

Oʻahu

Hawaiʻi is the 4th smallest State of the U.S. and is located 3,800 km West of the mainland in the Pacific Ocean (Mogil, 2013, 27; DBEDT, 2019a, 2). The State is over 28,000 km^2 in area and has eight main islands out of a total of 137: Hawaiʻi Island, Maui, Kahoʻolawe, Lānaʻi, Molokaʻi, Oʻahu, Kauaʻi, and Niʻihau (DBEDT, 2019a, 2). Hawaiʻi's rainfall can be quite variable due to tropical cyclones and cold fronts (Mogil, 2013) and Stidd and Leopold (1951, 28) provide data that shows Oʻahu's annual rainfall can vary between 450 to 6,200 mm in specified areas. The DBEDT (2019a, 6) also quote average temperatures between 21 to 29°C.

According to Jones (2015) and the DBEDT (2019b), Hawaiʻi's most significant economic sectors are tourism, defence, agriculture, manufacturing and the service industry; it has developed a growing US$92 billion-dollar GDP economy (BEA, 2019, 3).

As hinted earlier, Oʻahu is set in the wider geopolitical context of the State of Hawaiʻi – in the same way that Trinidad is a part of T&T. Hawaiʻi is being studied because:

- *Prior MSc. research*: Neale and Jardine (2015) looked at social housing PV potential through an indexing framework called the Geographic Insolation Potential and its Index. They indexed and mapped all the States of the U.S., Provinces and Territories of Canada, and Municipalities of T&T, and of the 79 locations, Hawaiʻi topped the list and was the only location to achieve the Index's benchmark performance.

However, Oʻahu is specifically looked at because:

- *The State's capital is in Oʻahu*: Oʻahu is home to the State's capital, Honolulu, so it is the centre of policymaking, business and State governance (World Population Review, 2019; DBEDT, 2019a).

- *Oʻahu is the most populated and urbanized*: Hawaiʻi is the most isolated population centre on the planet with an estimated 1.4 million residents, of which 64 to 89% live in its urbanscape which covers 10% of the State's land space (HCZMP, 1996, 101; Kaya & Yalcintas, 2010, 1364; BEA, 2019; DBEDT, 2019a, 3). The State has an estimated 530,000 housing units and over 450,000 households (Tian, 2014, 2; DBEDT, 2019a, 3). However, Oʻahu has nearly 70% of the State's population, 65% of its housing units, and 69% of its households (Tian, 2014, 2; World Population Review, 2019; DBEDT, 2019a).
- *Oʻahu is the State's energy gateway*: Hawaiʻi has no indigenous hydrocarbons so the State relies on costly imports (Alm, 2015). Almost 80% of Hawaiʻi's energy is imported as petroleum (State of Hawaii, 2019) and the imports are delivered to Oʻahu, refined, and then shipped to the other islands (U.S. EIA, 2018).
- *Oʻahu has the greatest power generation capacity*: The State's electricity is generated by two companies operating four utilities, that is, the Kauai Cooperative in Kauaʻi, and the HEI that services the rest of the archipelago; the HEI is made up of the parent company, the HECO on Oʻahu, and its subsidiaries, MECO that electrifies Maui, Lānaʻi and Molokaʻi, and the HECLO on Hawaiʻi Island (DBEDT, 2014). The HEI provides 95% of the State's electricity but this is dominated by the HECO's generation (Pintz & Morita, 2017, 26).
- *Oʻahu has the highest energy demand*: Oʻahu has the highest energy consumption of any of the islands under the HEI's jurisdiction (HNEI, 2008; GE Energy Consulting, 2015). Each of the Hawaiʻian islands' electricity grids is also independent so this means that their respective generation capacities need to meet their demand.
- *Oʻahu has the largest residential electricity customer-base*: Of the HEI's estimated 460,000 residential customers, Oʻahu has over 300,000 which means that the HECO has the largest customer-base (DBEDT, 2015, 2; HECO, 2019a).
- *Oʻahu (and by extension Hawaiʻi) has an extensive residential solar energy policy history*: Oʻahu's policymakers have implemented several different policies which supported and still support multiple dimensions of the island's solar energy growth (and renewable energy more broadly). Examples are RPSs, NEM as well as federal and state tax credits.
- *Oʻahu leads Hawaiʻian solar*: The HECO's customer-sited PV generation capacity (620 MW) and number of systems installed (54,722) is almost five times greater than any of the other utilities' (DBEDT, 2015, 18; HECO, 2019b). Lim (2011, 8) also shows Oʻahu's standout solar energy development where most of the State's proposed solar energy projects (12 out of 15) were in Oʻahu.
- *Oʻahu has the largest renewable energy industry*: The County of Honolulu, the designation for Oʻahu, has almost 62% of the State's green energy jobs across all the industries (Lim, 2011, 19). In the wider scope, Lim

(2011, 7 and 8) also presents supporting contexts for the industry where the HECO's recognized number of distributed renewable energy systems and installed capacity were the largest and Oʻahu had the most renewable energy projects (26 out of 66) for example.

Book overview: the energy transitions narrative

The research which informs this book was qualitative. The empirical chapters (Chapters 4 to 7) are based on structured interviews conducted with residents in Trinidad and Barbados, as well as semi-structured, key-informant interviews conducted with high-level policy officials in Trinidad and Oʻahu – which were analysed using inductive and deductive thematic content analyses respectively.

The residential interviews focused on cultural themes and those with the policy-influencers targeted institutional topics. Therefore, there was cultural data collected in Trinidad and Barbados and policy/institutional data collected in Trinidad and Oʻahu – which is a research design inherited from the doctoral thesis that the book is based on. Nevertheless, this primary data was also complemented by and triangulated with literature reviews on energy transitions, 'culture' as well as policy-related work on each case study.

Energy transitions are processes in time, so this means that the various developments involved are time sensitive. Therefore, retrospectively looking back at historical energy transitions can create a sequence of developments along a timeline. If one were to pull out some key characteristics that can be used to position the three case study islands next to each other in relation to their residential (solar) energy transition(s), an emergent theme would be that the islands seem to have had no significant residential solar energy being utilized for a period after which household SWHs entered, followed by distributed household PV, and, then, battery-integrated PV. This observation follows a technological transition (recalling that 'technology' is one of the key elements of interest).

In that regard, the book will present empirical work from Trinidad first, followed by Barbados and then Oʻahu: Trinidad is an example of an extreme and atypical small, tropical at the start of the technological timeline, Barbados is an intermediary example with its mainstreaming of SWHs, and Oʻahu is the furthest along due to its mainstreaming of PV. Figure 1.3 presents a graphic of the transitions narrative that the empirical chapters follow.

But before these empirical chapters however, Chapter 2 explores the nature of and theory behind energy transitions. It describes the MLP as the lead theory being used to approach energy transitions in these small, tropical islands. As the title of the book suggests, 'mainstreaming' is a central concept. 'Mainstreaming' in the MLP however was not an explicit theoretical concept so Chapter 2 will also give context to this.

'Culture' was a gap in the MLP's approach to energy transitions as well; its definition and role was not well-articulated in the work on the MLP and

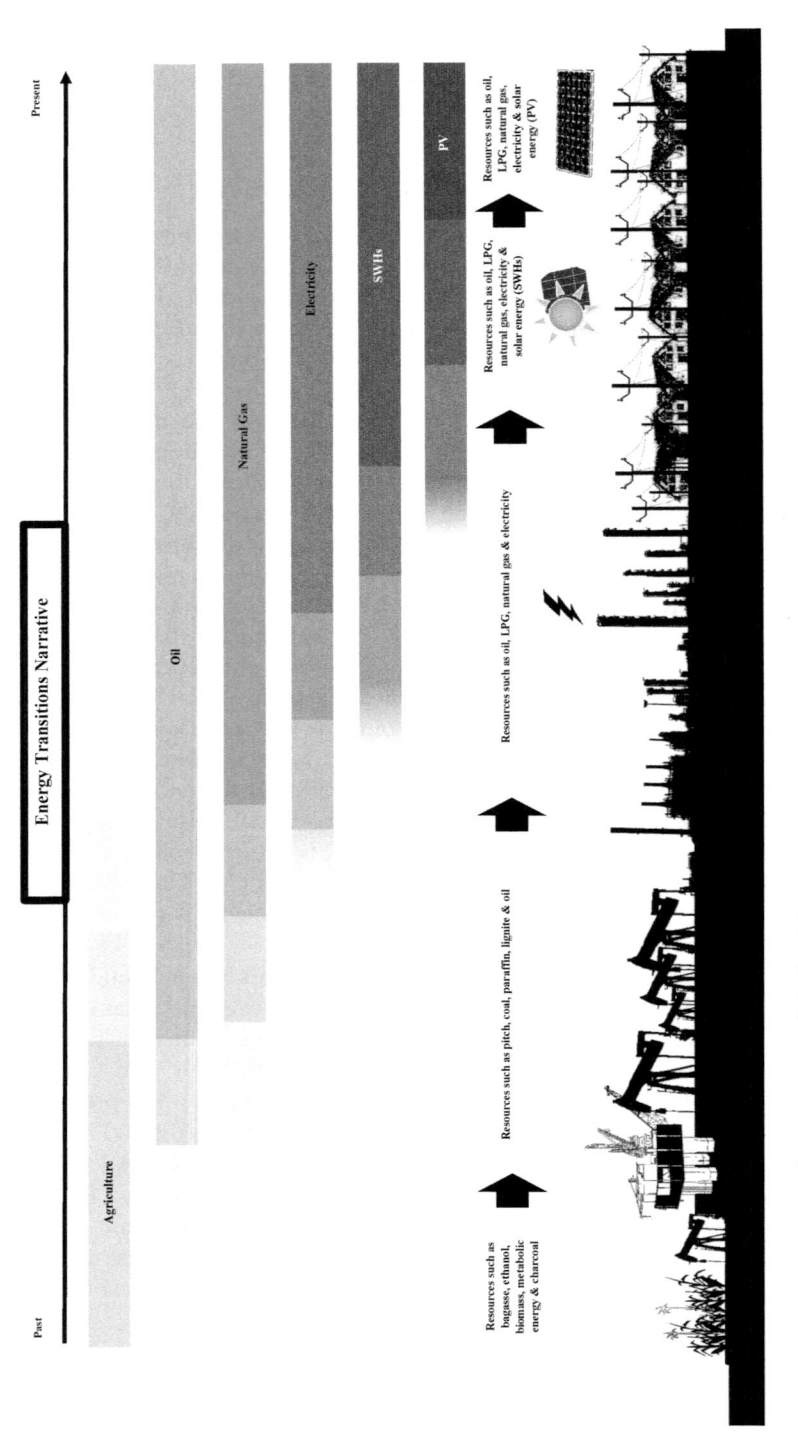

Figure 1.3 Showing a graphic summary of the book's energy transitions narrative.

'Sun, sea, sand' and solar energy   17

wider energy transitions literature which was reviewed. So, Chapter 3 broadly outlines what 'culture' means, as well as does so more specifically in relation to energy transitions by using the ECF (the book's second theory). Therefore, the book's theoretical framework is based on using the ECF to define and explain 'culture' in the MLP as an energy culture.

Chapter 4 is the first of the empirical chapters. It begins by looking at the earliest energy (re)sources in Trinidad, for example biomass and metabolic energy, and how these transitioned into the use of oil followed by natural gas with the dominance of electricity. The chapter is aimed at showing how the energy (re)sources of small, tropical islands' former agrarian economies transitioned into the fossil fuels and electricity of today. Further, the chapter outlines how residential solar energy fits into these transitions by looking at the state of solar energy implementation in Trinidad, for example the barriers and possible supporting policies.

Chapter 4 has an institutional and policy-oriented feel whilst Chapter 5 looks at the cultural side of this. It complements the above by exploring the cultural constructs of the energy supply, usage and payment systems of the incumbent residential electricity system in Trinidad. These two chapters therefore give a holistic impression of what the residential electricity system is like in a small, tropical island which has not mainstreamed household SWHs or PV.

Given that the role(s) of 'culture' was a gap in how the MLP approaches energy transitions and household SWHs seemed to have been mainstreamed before PV in the technological timeline of the case studies, Chapter 6 is dedicated to looking at Barbados' mainstreaming of solar water heating from a cultural angle. Chapter 6 also builds on the fact that electricity is the dominant energy carrier of today due to developments like those outlined in Chapters 4 and 5.

Like Chapter 5, the cultural aspects of the energy supply, usage and payments systems of Barbados' residential electricity system are looked at. This means that there is a subtle energy cultures comparison being made between Trinidad and Barbados in Chapter 6 – so doing will enable any differences attributable to the mainstreaming of solar water heating to stand out. It therefore gives insight into what an energy culture which has culturally assimilated solar energy is like. The chapter ends by touching on Barbados' transition from SWHs to/through distributed PV.

Chapter 7 picks up where Chapter 6 left off by outlining how Oʻahu transitioned from household SWHs to having mainstreamed distributed PV. This chapter takes an institutional and policy-oriented tone (like Chapter 5). It characterizes Oʻahu's residential electricity system and the impact of PV. However, it also and more prominently outlines how policies can be influential in mainstreaming solar energy. Chapter 7 ends by highlighting the policy developments linked to the transition from distributed PV to/through battery-integrated household PV.

The last chapter, Chapter 8, presents the conclusions of the research which informed the book's energy transitions narrative by focusing on the characterization of mainstreaming as a process; the role(s) of 'culture' in energy transitions based on its interactions with 'policy'; and the cultural and policy implications of and for mainstreaming solar energy.

References

Alm, R. (2015). *Hawaii's 100% Renewable Goal: The Confluence of Policy and Reality.* Paper presented at the International Conference on Perspectives on the Development of Energy and Mineral Resources Hawaii, Mongolia and Germany, Honolulu, Hawaii. Retrieved from http://socialsciences.hawaii.edu/conference/demr2015/_papers/alm-robert.pdf.

Arnesen, M. (2013). *Saving Energy Through Culture: A Multidisciplinary Model for Analyzing Energy Culture Applied to Norwegian Empirical Evidence* (Master's thesis, Trondheim, Norway: Norwegian University of Science and Technology). Retrieved from https://core.ac.uk/display/52098622.

Baek, S., Kim, H. & Chang, H. J. (2015). Optimal Hybrid Renewable Power System for an Emerging Island of South Korea: The Case of Yeongjong Island. *Sustainability*, 7(10): 13985–14001. Retrieved from https://doi.org/10.3390/su71013985.

Barron, J. & Sinnott, D. (2013). *Energy Use in Existing Dwellings: An Ethnographic Study of Energy Use Patterns in The Domestic Sector in Ireland.* Paper presented at CLIMA 2013–11th REHVA World Congress and the 8th International Conference on IAQVEC, Prague, Czech Republic. Retrieved from www.semanticscholar.org/paper/Energy-Use-in-Existing-Dwellings%3A-An-Ethnographic-Barron/c9408bfc71c68f00c12e002dcb6cac62c42c2206#citing-papers.

BEA (Bureau of Economic Analysis). (2019). *Hawaii.* United States Department of Commerce. Retrieved from www.bea.gov/regional/bearfacts/pdf.cfm?fips=15000&areatype=15000&geotype=3.

Beaie, S. T. (2009). *CARICOM Capacity Development Programme (CCDP) 2000 Round of Population and Housing Census Data Analysis Sub-Project National Census Report, Trinidad and Tobago.* Georgetown, Guyana: The CARICOM Secretariat. Retrieved from http://catalog.ihsn.org/index.php/catalog/4217/download/55709.

Bugler, W. (2012). *Seizing the Sunshine: Barbados' Thriving Solar Water Heater Industry.* Inside Stories on Climate Compatible Development, Climate & Knowledge Development Network and Acclimatise. Retrieved from https://cdkn.org/resource/cdkn-inside-story-seizing-the-sunshine-barbados-thriving-solar-water-heater-industry/?loclang=en_gb.

Cassen, C., Waisman, H., & Hourcade, J. C. (2012). *Sustainable Energy Transitions and the Economic Globalization.* Paper presented at the Sustainability Boundaries and the Great Transition, Planet Under Pressure Conference, London. Retrieved from https://halshs.archives-ouvertes.fr/halshs-00799522.

Castalia Limited. (2010). *Sustainable Energy Framework for Barbados ATN/OC-11473-BA. Final Report-Volume 1.* Government of Barbados and Inter-American Development Bank. Retrieved from https://bajan.files.wordpress.com/2011/07/barbados-sustainable-energy-framework-vol-i.pdf.

CDB (Caribbean Development Bank). (2014). *A New Paradigm for Caribbean Development: Transitioning to a Green Economy.* Barbados. Retrieved from www.caribank.org/

publications-and-resources/resource-library/thematic-papers/study-new-paradigm-caribbean-development-transitioning-green-economy.

CEIS (Caribbean Energy Information System). (2015, November 2015). *Solar Swells in Barbados: Capacity Set to Double in 2016.* Retrieved from www.ceis-caribenergy.org/solar-swells-in-barbados-capacity-set-to-double-in-2016-2/.

Cherry, T. L., Garcia, J. H., Kallbekken, S. & Torvanger, A. (2014). The Development and Deployment of Low-carbon Technologies: The Role of Economic Interests and Cultural Worldviews on Public Support. *Energy Policy, 68*: 562–566. Retrieved from https://doi.org/10.1016/j.enpol.2014.01.018.

Chivers, D. (2015). *Renewable Energy: Cleaner, Fairer Ways to Power the Planet.* Oxford, UK: New International Publications Limited.

Clkr-Free-Vector-Images. (2012a). [*Untitled vector of a sugarcane stalk*]. Pixabay. Retrieved from https://pixabay.com/vectors/sugarcane-cane-sugar-crop-42263/.

Clkr-Free-Vector-Images. (2012b). [*Untitled vector of an oil pump*]. Pixabay. Retrieved from https://pixabay.com/vectors/drilling-oil-rig-pump-petroleum-36265/.

Clkr-Free-Vector-Images. (2012c). [*Untitled vector of an offshore oil rig*]. Pixabay. Retrieved from https://pixabay.com/vectors/oil-rig-drilling-offshore-oil-29126/.

Clkr-Free-Vector-Images. (2014a). [*Untitled vector of a black plant*]. Pixabay. Retrieved from https://pixabay.com/vectors/black-silhouette-plants-white-294119/.

Clkr-Free-Vector-Images. (2014b). [*Untitled vector of a black lightning bolt*]. Pixabay. Retrieved from https://pixabay.com/vectors/bolt-lightning-silhouette-black-305692/.

CSO (Central Statistical Office). (2012). *Trinidad and Tobago 2011 Population and Housing Census Demographic Report.* Ministry of Planning and Sustainable Development, Government of the Republic of Trinidad and Tobago, Port of Spain, Trinidad and Tobago. Retrieved from www.tt.undp.org/content/dam/trinidad_tobago/docs/DemocraticGovernance/Publications/TandT_Demographic_Report_2011.pdf.

DBEDT (Department of Business, Economic Development and Tourism). (2015). *Hawaii Energy Facts and Figures May 2015.* State of Hawaii, Honolulu: Hawaii State Energy Office. Retrieved from http://energy.hawaii.gov/wp-content/uploads/2011/10/HSEO_FF_May2015.pdf.

DBEDT (Department of Business, Economic Development and Tourism). (2019a). *Hawai'i Facts and Figures. January 2019.* State of Hawaii, Honolulu: Hawaii State Energy Office. Retrieved from http://files.hawaii.gov/dbedt/economic/library/facts/Facts_Figures_browsable.pdf.

DBEDT (Department of Business, Economic Development and Tourism). (2019b). *What are the Major industries in the State of Hawaii?* State of Hawaii, Honolulu: Research and Economic Analysis. Retrieved from http://dbedt.hawaii.gov/economic/library/faq/faq08/.

DBEDT (Department of Business, Economic Development and Tourism). (2014). *Hawaii Energy Facts and Figures November 2014.* State of Hawaii, Honolulu: Hawaii State Energy Office. Retrieved from http://energy.hawaii.gov/wp-content/uploads/2014/11/HSEO_FF_Nov2014.pdf.

Dharshing, S. (2017). Household Dynamics of Technology Adoption: A Spatial Econometric Analysis of Residential Solar Photovoltaic (PV) Systems in Germany. *Energy Research & Social Science, 23*: 113–124. Retrieved from https://doi.org/10.1016/j.erss.2016.10.012.

Dirks, G., King, L., Laird, F., Lloyd, J., Lovering, J., Nordhaus, T., Pielke Jr., R., Román, M., Sarewitz, D., Shellenberger, M., Singh, K. & Trembath, A. (2014). *High-Energy Innovation: A Climate Pragmatism Project*. Oakland, CA: The Breakthrough Institute. Retrieved from https://thebreakthrough.org/articles/high-energy-innovation.

Dornan, M. (2015). Renewable Energy Development in Small Island Developing States of the Pacific. *Resources*, 4(3): 490–506. Retrieved from https://doi.org/10.3390/resources4030490.

Egmond, C., Jonkers, R. & Kok, G. (2006). One Size Fits All? Policy Instruments Should Fit the Segments of Target Groups. *Energy Policy*, 34(18): 3464–3474. Retrieved from https://doi.org/10.1016/j.enpol.2005.07.017.

Elias Sch. (2017). [*Untitled illustration of a small island in the Caribbean*]. Pixabay. Retrieved from https://pixabay.com/photos/island-vacations-caribbean-2482200/.

Espinasa, R. & Humpert, M. (2013). Trinidad and Tobago. In *Energy Matrix Country Briefings* (pp. 57–60). Inter-American Development Bank. Retrieved from https://publications.iadb.org/en/publication/11241/energy-matrix-country-briefings-antigua-barbuda-bahamas-barbados-dominica-grenada.

Espinasa, R. & Humpert, M. (2016). *Energy Dossier: Trinidad and Tobago*. Inter-American Development Bank. Retrieved from https://publications.iadb.org/en/energy-dossier-trinidad-and-tobago.

Everest-Phillips, M. (2014). *Small, So Simple? Complexity in Small Island Developing States*. Singapore: United Nations Development Programme Global Centre for Public Service Excellence. Retrieved from www.undp.org/content/dam/undp/library/capacity-development/English/Singapore%20Centre/GPCSE_Complexity%20in%20Small%20Island.pdf.

Fouquet, R. (2011). *Lessons from History for Transitions to a Low Carbon Economy*. Leioa, Spain: Public Policy Briefings 2011–04, Basque Centre for Climate Change (BC3). Retrieved from www.bc3research.org/policybriefings/2011-02.html.

Fouquet, R. & Pearson, P. J. G. (2012). Past and Prospective Energy Transitions: Insights from History. *Energy Policy*, 50: 1–7. Retrieved from https://doi.org/10.1016/j.enpol.2012.08.014.

Foxon, T. J., Hammond G. P. & Pearson, P. J. G. (2010). Developing Transition Pathways for a Low Carbon Electricity System in the UK. *Technological Forecasting and Social Change*, 77(8): 1203–1213. Retrieved from https://doi.org/10.1016/j.techfore.2010.04.002.

Fuchs, G. & Hinderer, N. (2016). One or Many Transitions: Local Electricity Experiments in Germany. *Innovation: The European Journal of Social Science Research*, 29(3): 320–336. Retrieved from https://doi.org/10.1080/13511610.2016.1188683.

GE (General Electric) Energy Consulting. (2015). *Technical Report: Hawaii Renewable Portfolio Standards Study*. Hawaii Natural Energy Institute. Retrieved from www.hnei.hawaii.edu/sites/www.hnei.hawaii.edu/files/Hawaii%20RPS%20Study%20-%20Final%20Report.pdf.

Geels, F. W. (2004). From Sectoral Systems of Innovation to Socio-technical Systems: Insights about Dynamics and Change from Sociology and Institutional Theory. *Research Policy*, 33(6–7): 897–920. Retrieved https://doi.org/10.1016/j.respol.2004.01.015.

Gischler, C., Medina, R., Ordóñez, J., Echevarría, C., Alleng, G., Beall, E., Medeazza, G. M.V., Franklin, R. & Buchara, D. (2009). *Sustainable Energy Framework for Barbados*

(BA-T1007) Plan of Operations. Inter-American Development Bank. Retrieved from www.energy.gov.bb/web/sustainable-energy-framework-for-barbados.

GORTT (Government of the Republic of Trinidad and Tobago). (2016). *Intended Nationally Determined Contribution (INDC) Under the United Nations Framework Convention on Climate Change: Executive Summary*. Port of Spain, Trinidad and Tobago: Multilateral Environmental Agreement Unit, Ministry of Planning and Development,.

GoB (Government of Barbados). (2013). *Barbados National Assessment Report: For the Third International Conference on Small Island Developing States, September 1–4, 2014 Apia, Samoa*. United Nations. Retrieved from https://sustainabledevelopment.un.org/content/documents/1054241Barbados_National_Assessment_Report_2014 August%20edition-2.pdf.

Government of Barbados (GoB). (2015). *National Sustainable Energy Policy (Revised)*. Retrieved from www.energy.gov.bb/web/component/docman/doc_download/58-nsep.

Gray, F., Koo, J., Chessin, E., Rickerson, W. & Curti, J. (2015). *Solar Water Heating Techscope Market Readiness Assessment. Reports for: Aruba, Bahamas, Barbados, Dominican Republic, Grenada, Jamaica, St. Lucia, Trinidad and Tobago*. Paris, France: Division of Technology, Industry and Economics, Global Solar Water Heating Initiative, United Nations Environment Programme. Retrieved from www.solarthermalworld.org/sites/gstec/files/story/2015-10-16/activity_5_-_swh_techscope_assessments_of_eight_caribbean_countries_report.pdf?ref_site=ppiaf.

Haraksingh, I. (2001). Renewable Energy Policy Development in the Caribbean. *Renewable Energy*, *24*(3): 647–655. Retrieved from https://doi.org/10.1016/S0960-1481(01)00051-9.

HCZMP (Hawaii Coastal Zone Management Program). (1996). Urban Areas. In *Hawaii's Coastal Nonpoint Pollution Control Program Management Plan* (Vol. 1) (pp. 101–159). Honolulu, State of Hawaii: Hawaii State Office of Planning. Retrieved from http://files.hawaii.gov/dbedt/op/czm/initiative/nonpoint/cnpcp_mgmt_plan.pdf.

HECO (Hawaiian Electric Company Incorporated). (2019a). *Power Facts*. Retrieved from www.hawaiianelectric.com/about-us/power-facts.

HECO (Hawaiian Electric Company Incorporated). (2019b). *Quarterly Installed Solar Data: 3rd Quarter, 2019*. Retrieved from www.hawaiianelectric.com/clean-energy-hawaii/our-clean-energy-portfolio/quarterly-installed-solar-data.

Hinchliffe, S. (1995). Missing Culture: Energy Efficiency and Lost Causes. *Energy Policy*, *23*(1): 93–95. Retrieved from https://doi.org/10.1016/0301-4215(95)90769-4.

HNEI (Hawaii Natural Energy Institute). (2008). *Assessment of the State of Hawaii's Ability to Achieve 2010 Renewable Portfolio Standards- Final Report*. University of Hawaii, Honolulu, State of Hawaii: School of Oceanography and Earth Science and Technology. Retrieved from http://puc.hawaii.gov/wp-content/uploads/2013/04/RTL-2010RenewablePortStd.pdf.

Husbands, J. (2016). *The History and Development of the Solar Hot Water Industry in Barbados*. Presentation published by olar Dynamics Limited. Retrieved from http://solardynamicslimited.com/wp-content/uploads/2016/10/Histor-Solar-Water-Heating-Industry-Barbados.pdf.

IEA (International Energy Agency). (2019). *Energy Security*. Energy Topics, Our Work. Retrieved from www.iea.org/topics/energysecurity/.

IEA and IRENA (International Energy Agency and International Renewable Energy Agency). (2017). *Perspectives for the Energy Transition: Investment Needs for a Low-Carbon Energy System*. Berlin, Germany: German Federal Ministry for Economic Affairs and Energy. Retrieved from www.irena.org/-/media/Files/IRENA/Agency/Publication/2017/Mar/Perspectives_for_the_Energy_Transition_2017.pdf?la=en&hash=56436956B74DBD22A9C6309ED76E3924A879D0C7.

IEA-ETSAP and IRENA (International Energy Agency Energy Technology Systems Analysis Programme and International Renewable Energy Agency). (2015). *Solar Heating and Cooling for Residential Applications Technology Brief*. Retrieved from www.solarthermalworld.org/sites/gstec/files/news/file/2015-02-27/irena-solar-heating-and-cooling-2015.pdf.

Ince, D. (2017). *Final Draft of the Energy Policy (2017–2037)*. Presented to the Division of Energy and Telecommunications, Prime Minister's Office, Government of Barbados, Barbados. Retrieved from www.energy.gov.bb/web/component/docman/doc_download/76-final-draft-of-national-energy-policy.

Ince, D. (2018). *Barbados National Energy Policy (2017–2037)*. Presented to the Division of Energy and Telecommunications, Prime Minister's Office, Government of Barbados, Barbados. Retrieved from www.energy.gov.bb/web/component/docman/doc_download/86-barbados-national-energy-policy-2017-2037.

Ioannidis, A. & Chalvatzis, K. J. (2017). Energy Supply Sustainability for Island Nations: A Study on 8 Global Islands. *Energy Procedia*, *142*: 3028–3034. Retrieved from https://doi.org/10.1016/j.egypro.2017.12.440.

Jackman, M. & T. Lorde, T. (2012). Examination of Economic Growth and Tourism in Barbados: Testing the Supply-Side Hypothesis. *Tourismos: An Internal Multidisciplinary Journal of Tourism*, *7*(2): 203–215. Retrieved from www.chios.aegean.gr/tourism/VOLUME_7_No2_art10.pdf.

Jensen, T. L. (2000). *Renewable Energy on Small Islands* (2nd ed.). Forum for Energy and Development. Retrieved from www.gdrc.org/oceans/Small-Islands-II.pdf.

Johnson Gordon. (2019). [*Untitled vector of a petrochemical factory*]. Pixabay. Retrieved from https://pixabay.com/vectors/petrochemical-factory-silhouette-4319118/.

Jones, S. (2015, March 3). Top 5 Industries in Hawaii: Which Parts of the Economy are the Strongest? *Newmax.com*. Retrieved from www.newsmax.com/FastFeatures/industries-hawaii-strongest-economy/2015/03/05/id/628087/.

Kabir, E., Kumar, P., Kumar, S., Adelodun A. A. & Kim, K. (2018). Solar Energy: Potential and Future Prospects. *Renewable and Sustainable Energy Reviews*, *82*: 894–900. Retrieved from https://doi.org/10.1016/j.rser.2017.09.094.

Kakazu, H. (2007). *Islands' Characteristics and Sustainability*. Paper presented at the Sasakawa Peace Foundation (SPF) Seminar on Self-supporting Economy in Micronesia. Retrieved from www.spf.org/yashinomi/pdf/pacific/economic/kakazu01.pdf.

Kaya, A. & Yalcintas, M. (2010). Energy Consumption Trends in Hawaii. *Energy*, *35*(3): 1363–1367. Retrieved from https://doi.org/10.1016/j.energy.2009.11.019.

Kemp, R. (1994). Technology and the Transition to Environmental Sustainability: The Problem of Technological Regime Shifts. *Futures*, *26*(10): 1023–1046. Retrieved from https://doi.org/10.1016/0016-3287(94)90071-X.

Khan, S. I. & Obaidullah, M. (2008). *Fundamentals of Solar Water Heaters*. Retrieved from http://citeseerx.ist.psu.edu/viewdoc/summary?doi=10.1.1.622.263.

Khan, Z. & Khan, A. A. (2017). Current Barriers to Renewable Energy Development in Trinidad and Tobago. *Strategic Planning for Energy and the Environment*, *36*(4): 8–23. Retrieved from https://doi.org/10.1080/10485236.2017.11863769.

King, R. (2009). Geography, Islands and Migration in an Era of Global Mobility. *Island Studies Journal*, *4*(1): 53–84. Retrieved from www.islandstudies.ca/sites/vre2.upei.ca.islandstudies.ca/files/ISJ-4-1-2009-RKing.pdf.

Lilienthal, P. (2007). High Penetration Renewable Energy for Island Grids. *Power Engineering*, 111: 90–96. Retrieved from http://homerenergy.com/pdf/High_Penetrations_of_Renewable_Energy_for_Island_Grids.pdf.

Lim, R. (2011). *DEBT Energy Plan*. Presentation given at the Hawaii Clean Energy Initiative Plenary Session, Honolulu, State of Hawaii. Retrieved from www.hawaiicleanenergyinitiative.org/storage/media/1-Lim.pdf.

Markard, L. & Truffer, B. (2008). Technological Innovation Systems and the Multi-level Perspective: Towards an Integrated Framework. *Research Policy*, *37*(4): 596–615. Retrieved from https://doi.org/10.1016/j.respol.2008.01.004.

Marzolf, N. C., Cañeque, F. C. & Loy, D. (2015). *A Unique Approach for Sustainable Energy in Trinidad and Tobago*. Washington, DC: Inter-American Development Bank. Retrieved from www.energy.gov.tt/wp-content/uploads/2016/08/A-Unique-Approach-for-Sustainable-Energy-in-Trinidad-and-Tobago.pdf.

MEEA (Ministry of Energy and Energy Affairs). (2012). *Renewable Energy and Energy Efficiency Policy Trends and Initiatives in T&T*. Government of the Republic of Trinidad and Tobago. Presented at the Third National CDM Capacity Building Workshop, Carlton Savannah Hotel, Port of Spain, Trinidad and Tobago: Government of the Republic of Trinidad and Tobago. Retrieved from http://trinidadandtobago.acp-cd4cdm.org/media/353735/re-ee-policy-trends-initiatives-tt.pdf.

MEWRD (Ministry of the Environment, Water Resources and Drainage). (2010). *Barbados National Assessment Report of Progress Made in Addressing Vulnerabilities of SIDS through IMPLEMENTATION of the Mauritius Strategy for Further Implementation (MSI) of the Barbados Programme of Action*. Government of Barbados, Barbados: Ministry of the Environment, Water Resources and Drainage. Retrieved from https://sustainabledevelopment.un.org/content/documents/1180barbados.pdf.

Miaschi, J. (2017, December 13). *How Many Islands Are There in the World?* Worldatlas.com. Retrieved from www.worldatlas.com/articles/how-many-islands-are-there-in-the-world.html.

Miller, C. A., Iles, A. & Jones, C. F. (2013). The Social Dimensions of Energy Transitions. *Science as Culture*, *22*(2): 135–148. Retrieved from https://doi.org/10.1080/09505431.2013.786989.

Mogil, H. M. (2013). The Weather and Climate of Hawaii. *Weatherwise*, *66*(6): 26–36. Retrieved from https://doi.org/10.1080/00431672.2013.839234.

Moore, W., Alleyne, F., Alleyne, Y., Blackman, K., Blenman, C., Carter, S., Cashman, A., Cumberbatch, J., Downes, A., Hoyte, H., Mahon, R., Mamingi, N., McConney, P., Pena, M., Roberts, S., Rogers, T., Sealy, S., Sinckler, T. & Singh, A. (2014). *Barbados' Green Economy Scoping Study*. Government of Barbados, University of West Indies – Cave Hill Campus, and United Nations Environment Programme, Barbados. Retrieved from www.un-page.org/file/1593/download?token=uiUKfl0J.

MPDE (Ministry of Physical Development Environment). (2001). *Barbados' First National Communications to the United Nations Framework Convention on Climate Change*. St. Michael, Barbados: Ministry of Physical Development Environment, Government of Barbados. Retrieved from http://unfccc.int/resource/docs/natc/barnc1.pdf.

MPSD (Ministry of Planning and Sustainable Development). (2014). *National Spatial Development Strategy for Trinidad and Tobago: Surveying the Scene- Background Information and Key Issues*. Trinidad and Tobago: Ministry of Planning and Sustainable Development, Government of the Republic of Trinidad and Tobago. Retrieved from www.planning.gov.tt/OurTnTOurFuture/documents/Surveying_the_Scene_web.pdf.

Mustapha, S. (2012). *Oil and Gas People and Government: Trinidad and Tobago's Experience*. Presentation given at the Revenue Management in Hydrocarbon Economics Conference, Port of Spain, Trinidad. Retrieved from https://sta.uwi.edu/conferences/12/revenue/documents/MsShakiraMustapha.pdf.

Narinesingh, D., Kumarsingh, K., Attzs, M., Gouveia, G., Beckles, D. M., Clarke, R., Chadee, D., Pantin, D. & Pollinais, S. (2013). *Second National Communication of the Republic of Trinidad and Tobago Under the United Nations Framework Conventional on Climate Change*. Trinidad and Tobago: Government of the Republic of Trinidad and Tobago. Retrieved from http://unfccc.int/resource/docs/natc/ttonc2.pdf.

NASA (National Aeronautics and Space Administration). (2019). *Prediction of Worldwide Energy Resources (POWER)* [Dataset]. NASA Langley Research Center (LaRC) POWER Project funded through the NASA Earth Science/Applied Science Program. Retrieved from https://power.larc.nasa.gov/data-access-viewer/.

Neale, K. & Jardine, C. (2015). The Geographic Insolation Potential and Trinidad and Tobago's Social Housing. *Caribbean Academy of Sciences E-Journal*, 8(1).

Neill, D. R. & Granborg, B. S. M. (1986). Report on the Photovoltaic R&D Program in Hawaii. Institute of Electrical and Electronics Engineers. *Transaction on Energy Conversion*, 1(4): 43–49. Retrieved from https://ieeexplore.ieee.org/document/4765773/metrics#metrics.

O'Connor, P. A. (2010). *Energy Transitions*. The Pardee Papers, No. 12. Boston, MA: The Frederick S. Pardee Center for the Study of the Longer-Range Future, Boston University. Retrieved from www.bu.edu/pardee/pardee-paper-012-energy/.

OpenClipart-Vectors. (2013a). [*Untitled vector world map outline*]. Pixabay. Retrieved from https://pixabay.com/vectors/world-map-continent-country-117174/.

OpenClipart-Vectors. (2013b). [*Untitled vector of power poles*]. Pixabay. Retrieved from https://pixabay.com/vectors/power-line-transmission-line-154729/.

OpenClipart-Vectors. (2013c). [*Untitled vector of the sun and a solar cell*]. Pixabay. Retrieved from https://pixabay.com/vectors/panel-solar-sun-solar-cell-energy-158630/.

OpenClipart-Vectors. (2016a). [*Untitled vector of a black house*]. Pixabay. Retrieved from https://pixabay.com/vectors/black-home-house-icon-white-1296170/.

OpenClipart-Vectors. (2016b). [*Untitled vector of a solar panel*]. Pixabay. Retrieved from https://pixabay.com/vectors/electricity-energy-green-panel-1298849/.

Otte, P. (2014). A (New) Cultural Turn Toward Solar Cooking – Evidence from Six Case Studies across India and Burkina Faso. *Energy Research & Social Science*, 2: 49–58. Retrieved from https://doi.org/10.1016/j.erss.2014.04.006.

Patwardhan, A., Azevedo, I., Foran, T., Patankar, M., Rao, A., Raven, R., Samaras, C., Smith, A., Verbong, G., Walawalkar, R., Panse, R., Ranjan, S. & Umarji N.

(2012). Transitions in Energy Systems. In T. B. Johansson, N. Nakicenovic & L. Gomez-Echeverri (Eds.), *Global Energy Assessment – Toward a Sustainable Future* (pp. 1173–1202). Cambridge, UK: Cambridge University Press and Laxenburg, Austria: The International Institute for Applied Systems Analysis. Retrieved from www.globalenergyassessment.org/.

Pintz, W. S. & Morita, H. (2017). Hawaii Policy Background. In *Clean Energy from the Earth, Wind and Sun: Learning from Hawaii's Search for a Renewable Energy Strategy* (1st ed.) (pp. 15–34). Cham, Switzerland: Springer International Publishing. Retrieved from https://doi.org/10.1007/978-3-319-48677-2.

REN21 (Renewable Energy Policy Network for the 21st Century). (2017). *Advancing the Global Renewable Energy Transition: Highlights of the REN21 Renewables 2017 Global Status Report in Perspective.* Paris, France: REN21 Secretariat, United Nations Environment. Retrieved from www.ren21.net/wp-content/uploads/2019/05/GSR2017_Highlights_English.pdf.

RIC (Regulated Industries Commission). (2016). *T&TEC's Annual Performance Indicator Report for the period July 2011 to June 2012. Information Document.* Retrieved from www.ric.org.tt/wp-content/uploads/2017/03/Performance-Indicator-Report-2011-2012_final-2016-12-29.pdf.

RIC (Regulated Industries Commission.). (2017). *T&TEC's Annual Performance Indicator Report for The Year 2015.* Retrieved from www.ric.org.tt/wp-content/uploads/2017/09/TTEC-Annual-Performance-Indicator-Report-2015-FINAL.pdf.

Rogers, T., Chmutina, K. & Moseley, L. L. (2012). The Potential of PV Installations in SIDS – An Example in the Island of Barbados. *Management of Environmental Quality, 23*(3): 284–290. Retrieved from https://doi.org/10.1108/14777831211217486.

Samuel, H. A. (2013). *A Review of the Status of the Interconnection of Distributed Renewables to the Grid in CARICOM Countries.* Caribbean Renewable Energy Development Programme, Caribbean Community and Deutsche Gesellschaft für Internationale Zusammenarbeit (GIZ) GmbH Germany, Castries, St. Lucia. Retrieved from www.scribd.com/document/252916035/CREDP-GIZ-A-Review-of-the-Status-of-the-Interconnection-of-Distributed-Renewables-to-the-Grid-in-CARICOM-countries-2013.

Sawyer, S. W. (1982). Leaders in Change: Solar Energy Owners and the Implications for Future Adoption Rates. *Technological Forecasting and Social Change, 21*(3): 201–211. Retrieved from https://doi.org/10.1016/0040-1625(82)90050-6.

Schlegelmilch, K. (2010). *Barbados: Options for Promoting Environmental Fiscal Reform in EC Development Cooperation. Country Case Studies of Barbados.* Forum Ökologisch-Soziale Marktwirtschaft, Berlin, Germany. Retrieved from www.foes.de/pdf/2010-09-29%20Barbados%20-%20FINAL.pdf.

SEforALL (Sustainable Energy for All). (2012). *Barbados: Rapid Assessment and Gap Analysis.* Retrieved from https://seforall.org/sites/default/files/Barbados_RAGA_EN.pdf.

Shirley, R. & Kammen, D. (2013). Renewable Energy Sector Development in the Caribbean: Current Trends and Lessons from History. *Energy Policy, 57*: 244–252. Retrieved from https://doi.org/10.1016/j.enpol.2013.01.049.

Solaun, K., Gomez, I., Larrea, I., Sopelana, A., Ares, Z. & Blyth, A. (2015). *Strategy for Reduction of Carbon Emissions in Trinidad and Tobago, 2040: Action Plan for the*

Mitigation of GHG Emissions in the Electrical Power Generation, Transport and Industry Sectors. Port of Spain, Trinidad and Tobago: Government of the Republic of Trinidad and Tobago.

Sovacool, B. K. & Geels, F. W. (2016). Further Reflections on The Temporality of Energy Transitions: A Response to Critics. *Energy Research & Social Science*, *22*: 232–237. Retrieved from https://doi.org/10.1016/j.erss.2016.08.013.

State of Hawaii. (2019). *About the Hawaii Clean Energy Initiative*. Hawaii Clean Energy Initiative, State of Hawaii and United States Department of Energy. Retrieved from www.hawaiicleanenergyinitiative.org/about-the-hawaii-clean-energy-initiative/.

Statistics Section. (2019). *Trinidad and Tobago Country Fact Sheet*. Department of Foreign Affairs and Trade, Australian Government. Retrieved from http://dfat.gov.au/trade/resources/Documents/trin.pdf.

Stephenson, J. (2018). Sustainability Cultures and Energy Research: An Actor-centred Interpretation of Cultural Theory. *Energy Research & Social Science*, *44*: 242–249. Retrieved from https://doi.org/10.1016/j.erss.2018.05.034.

Stephenson, J., Barton, B., Carrington, G., Gnoth, D., Lawson, R. & Thorsnes, P. (2010). Energy Cultures: A Framework for Understanding Energy Behaviours. *Energy Policy*, *38*(10): 6120–6129. Retrieved from https://doi.org/10.1016/j.enpol.2010.05.069.

Stidd C. K. & Leopold, L. B. (1951). The Geographic Distribution of Average Monthly Distribution Rainfall, Hawaii. In: On the Rainfall of Hawaii. *Meteorological Monographs*, *1*(3): 24–33. American Meteorological Society, Boston, MA. Retrieved from https://doi.org/10.1007/978-1-940033-01-3_3.

Strauss, S., Rupp, S. & Love, T. (2013). Introduction. Powerlines: Cultures of Energy in the Twenty-first Century. In S. Strauss, S. Rupp & T. Love (Eds.), *Cultures of Energy: Power, Practices, Technologies* (1st ed.) (pp. 10–38). Walnut Creek, CA: Left Coast Press. Retrieved from https://doi.org/10.4324/9781315430850.

Szabó, S., Kougias, I., Moner-Girona, M. & Bódis, K. (2015). Sustainable Energy Portfolios for Small Island States. *Sustainability*, *7*(9): 12340–12358. Retrieved from https://doi.org/10.3390/su70912340.

Thomas, A. (2012, March 1). Frieburg Solar Region. *Word Wide Fund for Nature*. Retrieved from http://wwf.panda.org/?204416/Freiburg-Solar-Region.

Thomas, S. M. (2009). *Impacts of Economic Growth on CO_2 Emissions: Trinidad Case Study*. Paper presented at the 45th ISOCARP Congress, Porto, Portugal. Retrieved from www.isocarp.net/Data/case_studies/1598.pdf.

Thompson, E. H. (2015). *Lessons from the Reform of the Barbados Energy Sector*. Presentation published by the Energy Sector Management Assistance Program for Small Island Developing States, The World Bank, Washington, DC. Retrieved from https://esmap.org/sites/esmap.org/files/DocumentLibrary/3a%20-%20Elizabeth%20WB%20ESMAP%20PP_Optimized.pdf.

Tian, E. (2014). *What Are the Economic Drivers for Hawaii in 2014 and Beyond*. Presentation given at the Honolulu Board of Realtors Housing Forum, Honolulu, State of Hawaii. Retrieved from http://files.hawaii.gov/dbedt/economic/reports/2014-economic-drivers.pdf.

Truffer, B. & Coenen, L. (2012). Environmental Innovation and Sustainability Transitions in Regional Studies. *Regional Studies*, *46*(1): 1–21. Retrieved from https://doi.org/10.1080/00343404.2012.646164.

U.S. (United States) Department of Energy. (2004). *The History of Solar*. Retrieved from www1.eere.energy.gov/solar/pdfs/solar_timeline.pdf.

U.S. EIA (United States Energy Information Administration). (2018). *Hawaii State Profile and Energy Estimates: Profile Analysis*. Retrieved from www.eia.gov/state/analysis.cfm?sid=HI.

UN-OHRLLS (Office of the High Representative for the Least Developed Countries, Landlocked Developing Countries and Small Island Developing States). (2011). *Small Island Developing States: Small Islands Big(ger) States*. United Nations, New York. Retrieved from http://unohrlls.org/custom-content/uploads/2013/08/SIDS-Small-Islands-Bigger-Stakes.pdf.

van de Vijver, F. & Leung, K. (1997). Methods and Design. In *Methods and Data Analysis for Cross-Cultural Research* (Vol. 1) (pp. 27–79). Thousand Oaks, CA: SAGE Publications. Retrieved from https://us.sagepub.com/en-us/nam/methods-and-data-analysis-for-cross-cultural-research/book6080.

Verdolini, E. & Galeotti, M. (2011). At Home and Abroad: An Empirical Analysis of Innovation and Diffusion in Energy Technologies. *Journal of Environmental Economics and Management, 61*(2): 119–134. Retrieved from https://doi.org/10.1016/j.jeem.2010.08.004.

WEC (World Energy Council). (2016). *World Energy Resources- Solar 2016* (pp. 26–27). London. Retrieved from www.worldenergy.org/assets/images/imported/2016/10/World-Energy-Resources-Full-report-2016.10.03.pdf.

Wellington, P. (2011). *Meteorological Hazards Affecting Trinidad and Tobago: Disaster Risk Reduction*. Trinidad and Tobago: Trinidad and Tobago Meteorological Services (TTMS) and Committee for Meteorological Enhancement in Trinidad and Tobago (COMET), Piarco. Retrieved from www.metoffice.gov.tt/sites/default/files/Risk%20Reduction%20Handout.pdf.

Wolf, F., Surroop, D., Singh, A. & Leal, W. (2016). Energy Access and Security Strategies in Small Island Developing States. *Energy Policy, 98*: 663–673. Retrieved from https://doi.org/10.1016/j.enpol.2016.04.020.

Wong, P. P. (2011). Small Island Developing States. *Wiley Interdisciplinary Reviews: Climate Change, 2*(1): 1–6. Retrieved from https://doi.org/10.1002/wcc.84.

Wong, P. P., Marone, E., Lana, P., Fortes, M., Moro, D., Agard, J., Vicente, L., Thönell, J., Deda, P. & Mulongoy, K. J. (2005). Island Systems. In R. Hassan, R. Scholes & N. Ash (Eds.), *Ecosystems and Human Well-being: Current State and Trends* (Vol. 1) (pp. 663–680). Millennium Ecosystem Assessment, United Nations Environment Programme. Washington, DC: Island Press. Retrieved from www.millenniumassessment.org/documents/document.292.aspx.pdf.

World Bank (2019). *Gross Domestic Product 2018*. World Development Indicators Database. Retrieved from http://databank.worldbank.org/data/download/GDP.pdf.

World Population Review. (2019). *Hawaii Population 2019 (Demographics, Maps, Graphs)*. Retrieved from http://worldpopulationreview.com/states/hawaii-population/.

2 Energy transitions and the mainstream

Introduction

Without using energy, the quality of life cannot be much higher than that of primitive man (McVeigh, 1984). But no living organism produces more energy than it uses or produces any by itself because all are dependent on external fuel sources (Crosby, 2006). Today's energy intensive consumption depends on the extraction of external fuel sources like fossil fuels (Horta et al., 2014; Urry, 2014).

But the way in which these resources have been historically used shows that today's consumption and socio-economic systems are unsustainable (Pérez-Lombard, Ortiz & Pout, 2008). This also extends to the underlying energy system(s). In that regard, Westphal (2012) states that enhancing the energy system's sustainability requires the integration of renewables and the wider sustainable energy transitions literature suggests that renewables will eventually be a part of future energy systems (Steg, Perlaviciute & van der Werff, 2015) because the world is transforming how it accesses and uses energy (Manning, 2015).

Energy systems are complex, central to any economy, and made up of multiple sectors, institutions and resources (WEF, 2018). But having used the term 'transition' above, the term suggests that systems-level changes are not sudden and are rather prolonged processes (WEF, 2013). This consequently requires considering what constitutes an energy transition.

It is important to firstly acknowledge that humanity craved solar energy in one form or another (Crosby, 2006), and there have been many forms over humanity's energy history. In that regard, authors have tended to characterize energy transition chronologies through technological developments and usage of energy types (Gismondi, 2018).

The world's energy transition history involved the metabolic energy of human and animal labour as well as biomass being phased out for coal, followed by today's oil, natural gas and electricity (Grubler, 2012a, 2012b; WEF, 2013; Pain, 2017; Buchanan, 2017). The WEF (2013, 6) illustrates this as the biomass and coal contributions to the global energy mix decreased from the 1800s and 1900s respectively, and those of oil, gas and forms of electricity increased.

The last is a quintessential part of modern transitions (Smil, 2016) because it changed the way the world related to energy and became connected through energy (Namias, 2008; Love & Garwood, 2013; WEF, 2013).

Today's global energy is dominated by fossil fuels (European Commission, 2011; Smil, 2016) and they were able to become the dominant global energy resources within a half-century of their introduction (Kemp, 1994). However, renewable energy is the next envisioned transition paradigm (Herrera, 2015). Brinkworth (1985), Westphal (2012), the IEA and IRENA (2017) and Grayson (2017) suggest that though the transition from fossil fuel energy will take a long time, the process has already been set in motion.

The REN21 (2017) illustrates the trend in the growth in renewables and specific examples are the global PV and wind energy capacities growing from 6 to 303 GW and from 74 to 487 GW respectively between 2006 and 2016. Such trends are suggestive of a global energy transition unfolding. However, Elias and Victor (2005) and O'Connor and Cleveland (2014) outline that defining and, by extension, quantifying energy transitions is difficult. Sovacool (2016) further states that there is no universal definition for it thus far.

O'Connor and Cleveland (2014) suggest that one reason is because energy systems are dynamic structures. Cherp et al. (2018) suggest that the diversity of disciplinary approaches available to study energy systems is another. Also tied to these notions are the terminologies and scopes used since these vary together with their methods of analysis, for example, 'energy transition' versus 'low-carbon transition' or 'sustainable transition' for instance (Cherp et al., 2018).

O'Connor (2010) outlines that energy transitions affect energy resources, carriers, converters and services. They are commonly characterized via primary energy supply quantifications (Miller et al., 2013; Herrera, 2015; Smil, 2016; Fouquet, 2016), total final energy consumption, final energy intensity, electricity consumption or electricity and fuel prices (Hauff et al., 2014).

Despite such a variety of metrics, the IEA and IRENA (2017) state that the definition of energy transitions needs to transcend both energy supply and demand; however, Grubler (2012a) and Fouquet and Pearson (2012) state that the understanding of energy transitions has been too heavily weighted by the supply side.

A starting solution to this would be approaching energy transitions as socio-technical systems (Palm, 2006; Miller et al., 2013). These are systems of knowledge, practices and networks linked to energy technologies (Cherp et al. 2018) which develop over decades and tend to exhibit a path dependency that resists change (Kuzemko et al., 2016; Geels, 2018); the 'socio' relates to the social elements and the 'technical' to the technological components.

Socio-technical energy transitions can be thought of as large-scale societal transformations which fundamentally change the structure and nature of the socio-technical energy system (Rhodes, 2007; Kuzemko et al., 2016). Energy transitions are socio-technical because without the 'socio', it is not a complete systems-level transition and elements that inherently shape the transition run the risk of not being duly considered, for example, people's beliefs that drive their energy use which in turn affects the technical requirements of the energy system.

Based on this, an energy transition is a fundamental change that affects either the supply or demand sides of an energy system, and which has implications for both once the change occurs – and this is where the MLP comes in.

Characterizing the multi-level perspective

Introducing the multi-level perspective of socio-technical energy transitions

The book is an energy transitions study set within the socio-technical scholarship that spans evolutionary economics; science and technological studies; ecology; ecological economics; political science; history; institutional theory; and environmental studies (Lawhon & Murphy, 2011; Sovacool & Hess, 2017). There are many existing theories, for example, diffusion of innovations, path dependency and actor-network theories, and Sovacool and Hess (2017) present a useful overview of 14 of them.

The book however uses the MLP:.this is a vertically structured theory built on temporal and structural scales (Raven, Schot & Berkhout, 2012; Hargreaves, Longhurst & Seyfang, 2013), and draws on the historical and sociological exploration of technology largely through explanations based on processes and contexts (Geels, 2018). The MLP is used because it adequately examines and simplifies complex processes (Jenkins, Sovacool & McCauley, 2018) and provides a relatively simple but highly adaptable framework for thinking about energy transitions (Hargreaves, Longhurst & Seyfang, 2013).

Figure 2.1 shows the MLP and it is one of the most significant conceptual approaches to socio-technical energy transitions (Kuzemko et al., 2016; Jenkins, Sovacool & McCauley, 2018). For instance, in Sovacool and Hess (2017), of the 96 theories covering 22 disciplines that arose from their work related to the theories commonly used to approach socio-technical change, the MLP was the most frequently cited. The MLP provides a systems-level context for energy transitions by looking at interactions between a socio-technical landscape, socio-technical regime and niche-innovations. Each will now be looked at in more detail.

The landscape, regime and niche-innovations

The landscape plane of the MLP is the most structured (Geels & Schot, 2007). Geels and Schot (2007), Kern and Smith (2008), Raven, Schot and Berkhout (2012) and Lockwood et al. (2013) characterize it as an exogenous plane that indirectly exerts change-pressures on the regime and niches over prolonged periods of time. These change-pressures create opportunities for niche-innovations to enter the regime depending on the latter's response to these pressures (Geels & Schot, 2007; Smith, Voß & Grin, 2010; Lockwood et al., 2013; Mattes, Huber & Koehrsen, 2015).

Landscape-forces are long-term structures (Markard & Truffer, 2008) capable of changing over time (Raven, Schot & Berkhout, 2012) and examples are oil prices, wars, major discoveries and wider social trends (Truffer & Coenen, 2012). The downward, dotted arrows from the landscape to the regime and from the landscape through the regime to niches in Figure 2.1 illustrate its influence.

Smith, Voß and Grin (2010) define regimes as structures made up of a co-evolutionary alignment of features. Geels and Schot (2007) specify that it is an alignment between science, policies, markets, industries, technologies and cultures, as well as the various actors within these spaces (see Figure 2.1). Examples of regime structures are centralized power generation plants and electricity grid networks.

Relative to the regime's actors, there are multiple which share dynamic inter-relationships (Markard & Truffer, 2008; Lockwood et al., 2013), so they are not isolated (Cressman, 2009; Nimmo, 2011; Pollack, Costello & Sankaran, 2013).

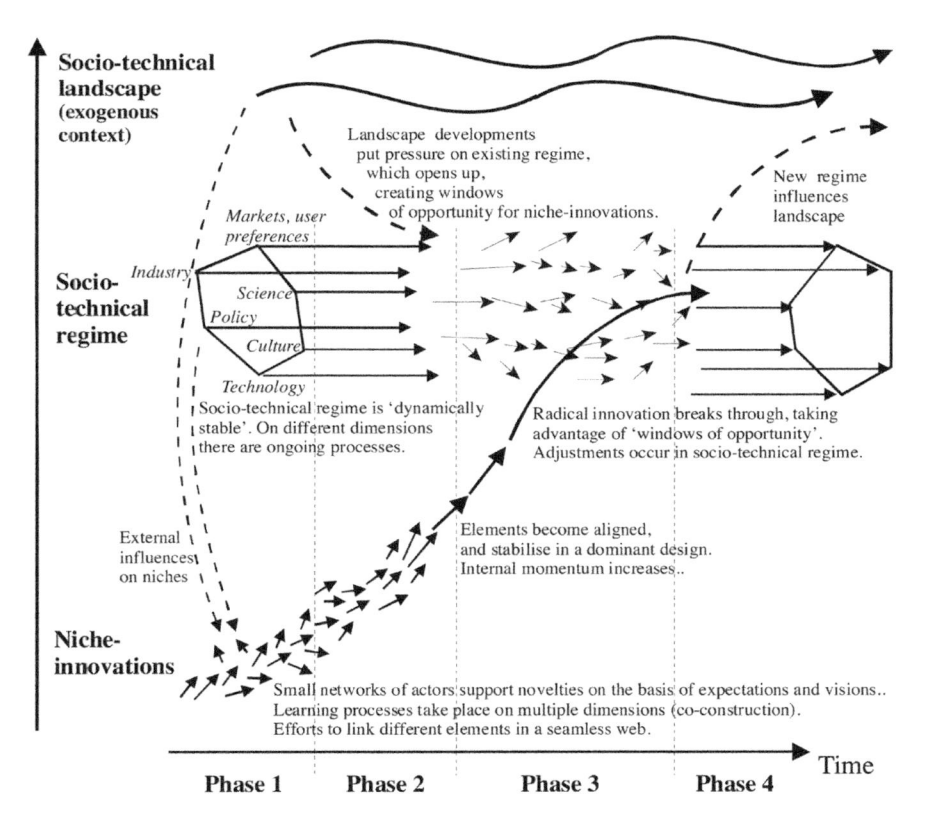

Figure 2.1 Showing the phases of an energy transition in the MLP.

The regime includes a multi-actor network constructed from a series of financial, supplier, user, producer, research, societal and public actor-clusters whose functions, that is, their agency, interact at the socio-technical level (Geels, 2002, 2004). The network is therefore influenced by actors' agency but also provides the exhibition space for their agency (Brass & Burkhart, 1993; Geels, 2004).

Cherp et al. (2018) outline that regimes exert an inertia which resists change. However, Markard and Truffer (2008) suggest that regimes are not completely resistant to change. Over time they are faced with new change-pressures (Schot, Hoogma & Elzen, 1994; Smith, Voß & Grin, 2010) so their responses to changes not only vary depending on the change factors but also the evolutionary nature of those factors (Smith, Voß & Grin, 2010; Truffer & Coenen, 2012).

Smith, Stirling and Berkhout (2005) add that the more disruptive change-influences are, the more resistant to change the regime is since there is a greater allocation of resources dedicated to maintaining its incumbency. Regimes therefore exhibit an adaptive capacity by responding to influences and is one reason why niche-innovations face challenges during energy transitions (Smith, Stirling & Berkhout, 2005; Markard & Truffer, 2008; Smith, Voß & Grin, 2010; Cherp et al., 2018).

Geels and Schot (2007) and Hargreaves, Longhurst and Seyfang (2013) suggest that an energy transition can be thought of as a shift from one socio-technical regime to another in the MLP. Therefore, Figure 2.1's left polygon evolving into the one on the right means that a transition is suggestive of a movement from one state to another (Sarrica et al., 2016).

Shifts in regime-states are often thought to be incremental (Mattes et al., 2015). But Smith, Stirling and Berkhout (2005) argue that if the resources needed to stimulate a regime-change are present internally, only then would the transition be incremental and non-disruptive – and incremental development is part of the nature of technological and mainstream changes (Kemp, 1994; Smith, Voß & Grin, 2010; Winskel et al., 2014; Klitkou et al., 2015) which is important given the title of this book.

A caveat is that such incremental development has the possible risk of the incoming structures being seamlessly encapsulated into the incumbent (Kern & Smith, 2008). This is because a regime supports niche-innovations that display characteristics which align with its incumbent orientation (Strebel, 2011) – bearing in mind that the technologies or energies which have become dominant today have had their origins in niches (Fouquet, 2016).

Niches are the least aggregated level of the MLP (Markard & Truffer, 2008). Smith, Voß and Grin (2010) and Grubler (2012a) describe niches as early trials and learning spaces. They harbour the novelties of transitions, are advocated for by small social networks and offer some degree of protection for innovations that are not yet competitive enough to make it into the regime (Geels & Schot, 2007; Smith, Voß & Grin, 2010; Fouquet, 2016; Cherp et al., 2018).

Niches are successful when they dissolve (Schot, 1998) because their innovations no longer need protection from mainstream forces and can compete with the incumbent regime (Yanosek, 2012). When this occurs, an innovation becomes part of the mainstream because it is more widely adopted (Berkhout, Smith & Stirling, 2004). Examples of niche-level elements are innovation policies for early adopters; lead markets; incentivizing the provisioning of cheaper services; subsidized demonstrations; and experimentation projects (Smith, Voß & Grin, 2010; Fouquet, 2011; Green, Skerlos & Winebrake, 2014).

Geels (2002, 2004) and Kuzemko et al. (2016) identify with the need for innovation-protection through the above examples because innovations are usually expensive and have low technical performances; innovations supported by protective mechanisms often have better chances of success (Kemp, 1994; Green, Skerlos & Winebrake, 2014).

Innovations can be product-related, process-oriented or managerial, and exist technologically, psychologically or even socio-culturally (Daghfous, Petrof & Pons, 1999; Zailani et al., 2015). However, energy transitions show a trend toward cheaper, cleaner, more abundant and reliable fuels which is complemented by the replacement of older energy-conversion technologies with newer options (Dirks et al., 2014). But it is for this reason that the MLP's applications and wider energy transitions literature has been criticized for being technologically biased (Hargreaves et al., 2013; Fuchs & Hinderer, 2016) and focused on technological substitutions (Geels, 2006).

Nevertheless, developing innovations is a non-linear process (Dirks et al., 2014), as are energy transitions (Geels & Schot, 2007; Fouquet, 2016; Andrews-Speed, 2016). Further, developing innovations is like the mainstream and technological changes that are part of energy transitions in that it is incremental, cumulative and assimilative (Beerepoot & Beerepoot, 2007).

When a niche-innovation enters the regime, it reconfigures the incumbent regime. This reconfiguration realigns heterogeneous attributes (Geels, 2002) and should be considered in light of alignment suggesting stability (Greenhalgh & Stones, 2010). The annotation just beneath Figure 2.1's left regime-polygon shows this as dynamic stability, and the annotation in the middle-right of the cascade of upward arrows between the niche and regime planes captures the realignment-stability relationship.

In summary, Geels (2006, 2018) outlines four key processes unfolding in the MLP, but then lays them out in Figure 2.1's phases: innovations emerge in Phase 1; niches are stabilized and innovations enter small niches in Phase 2; innovations break into the regime in Phase 3 through the landscape's influence and supporting developments in the niches and regime; and Phase 4 marks the regime's transformation. But there are several assumptions and limitations associated with the MLP worth pointing out.

Three main assumptions are: there are only three planes operating during energy transitions, that is, niche-innovations, the socio-technical regime and the socio-technical landscape; an energy transition unfolds when landscape-forces affect the regime to create the space(s) for niche-innovations to enter and structurally re-orient the incumbent regime; and landscape-forces take longer than the regime's to influence energy transition processes whilst the regime's take longer than niches', and landscape and regime-forces unfold over decadal to century timeframes whilst niche-level processes unfold within years.

Two main limitations are: the three planes are geographically fluid in their interpretation and there is no concrete boundary distinguishing one plane from the other, that is, the MLP falls short on spatial considerations as it relates to the conceptual geographic extents of niches, regimes and the landscape as well as the way in which space and place influence and are influenced by transitions; and the MLP's application in the literature is largely technologically oriented though the MLP is not technologically biased.

Energy transition timeframes and small islands

The incumbent global energy system based on fossil fuels includes the largest and most expensive anthropogenic infrastructure so even if alternative energy resources are more widely available, the scale of the incumbent system should not be taken for granted (Smil, 2016). This means that the legacy of fossil fuels has given its regime(s) inherent structural advantages over incoming renewable alternatives (Dumas et al., 2016).

However, where fossil energy has not been fully accessed or adopted, its lock-in effect has been limited and suggests that renewables can rapidly take off (Bridge et al., 2013). But based on Lacey's (2010) conversation with Vaclav Smil, the amount of today's hydrocarbon energy consumption that needs to be substituted by renewables amongst other frictional factors would mean that the transition to renewables is a prolonged one and not sudden and revolutionary.

There are many studies that take an optimistic outlook towards how fast transitions can unfold (Grubler, 2012a). But these outlooks should be considered in light of Fouquet and Pearson's (2012) analysis of historical transitions that shows that market size has been a determinant of how fast it can occur because as the size of the economy increases, so too does the length of time taken to transition.

There is historical evidence to support the prolonged timeframes of energy transitions (Sovacool, 2016) and the wider academic energy transitions literature focus on them as prolonged phenomena (Sovacool & Geels, 2016); for example Lawhon and Murphy (2011), Manning (2015), and Gates' (2015) work that state that energy transitions unfold over decades or longer. Fouquet and Pearson (2012) outline that from their work there is a 40-year minimum over which an energy transition unfolds. Raven, Schot and Berkhout (2012) further state that niche-level developments may unfold within ten years, regimes within 50 and landscape developments over centuries.

The way in which an energy transition is defined, however, will affect its timescales of operation (Sovacool, 2016; Fouquet, 2016) because there can be transitions that make up the holistic energy transition (Markard & Truffer, 2008; Fouquet, 2016). For example, Sovacool (2016) presents empirical evidence of faster transitions occurring in end-use devices and prime movers:

lighting in Sweden, cooking stoves in China, air conditioning in the U.S., and LPG in Indonesia.

Sovacool also gives further examples of rapid supply-oriented transitions through nuclear electricity in France, oil and electricity in Kuwait, and natural gas in the Netherlands. These examples lean on the definition of an energy transition as a sub-system unit under analysis, for example end-use technology, prime movers or energy supply versus the entire energy system.

Defining energy transitions in the MLP relates to defining the regime. The energy transitions literature focuses on national-scale geographies and conceptual developments where niches have often been framed as local developments, regimes as national and the landscape as international (Raven, Schot & Berkhout, 2012). For example, Hauff et al. (2014) define an energy transition as a fundamental change in the energy system of a country – country being the geographically relevant notion that speaks to the national scale.

Energy transitions theories do not capture sub-national dynamics (Hiemstra-van der Horst & Hovorka, 2008) and largely overlook the role of local processes (Mattes, Huber & Koehrsen, 2015); the MLP has likewise been criticized for being short on spatial considerations (Lawhon & Murphy, 2011; Truffer & Coenen, 2012; Truffer, Murphy & Raven, 2015; Mattes, Huber & Koehrsen, 2015).

Further, Smith, Voß and Grin (2010) and Raven, Schot and Berkhout (2012) describe the regime as being geographically fluid in its interpretation (since the MLP's planes are not geographically bound). However, Markard and Truffer (2008) state that defining the regime should be done based on using its conceptual strengths most effectively.

In the context of a small, tropical island, its geographic insularity acts as an automatic and strong framing factor for conceptualizing the regime. The book's design, that is, using a three-island narrative, also calls for an intra-island definition of the regime. So relative to time, by defining the regime as the residential electricity sector on an island, two elements stand out.

The first is related to the specificity of looking at small, tropical islands. Using such islands as the case geographies means that an energy transition would unfold faster here than in larger geographies; an energy transition should unfold faster in Trinidad, Barbados and/or Oʻahu than it would in Japan or the United Kingdom, for example.

The second is the socio-technical specificity of the residential electricity regime. This means that at the intra-island level, an energy transition should unfold faster here than if the regime was approached as the economy-wide energy system of the island which would include other sectors such as commercial and industrial sectors, and systems beyond electricity systems such as the transport fuel sector.

These other sectors would include different elements such as non-residential electricity tariffs; markets based on economies of scale; diesel and gasoline infrastructure; and vehicles which would change the nature of the regime being defined. In this regard, it is worth highlighting some of the motivations for conceptually looking at the residential electricity regime:

- *Energy and shelter are basic human needs*: All living things need energy to survive and all anthropogenic activity is connected to energy (Eni, 2016). The household is also the general framework in which most individuals identify with since the majority live in households and it is made up of persons living together, that is, sleeping and sharing meals together (CSO, 2012). So, based on these notions there is an anthropological link between consuming energy and sheltered living.

  This is more interestingly set in the context of households traditionally being consumers of energy. But with the adoption of household PV for instance, households can become prosumers, that is, consumers and producers of electricity (Eid et al., 2014; Georges et al., 2014) – which represents a fundamental change in households' role in an energy system.
- *The residential sector is a key electricity consuming sector*: For example, the residential electricity sector consumes roughly 29% of the electricity in T&T (Espinasa & Humpert, 2016, 14); 33% in Barbados (Ince, 2018, 42); and 28% in Hawai'i (READ, 2019).
- *The residential sector has more electricity customers than other sectors*: Each household legally connected to the grid is a utility company's customer. This means that the sheer number of households suggests that this sector is likely to have the largest number of customers.
- *The residential sector has more potential IPPs than other sectors*: Individual households tend to make up the largest portion of a utility company's customer-base. This means that relative to power generation and the centralized versus decentralized dichotomy, households likely to be distributed power generation agents will potentially account for the largest number of IPPs. This would especially be relevant if grid-connected policies are considered where households can generate, export and sell their solar power to utility companies.
- *Distributed solar energy is more likely given the limits of land area in small islands*: Small islands have limited land space. This means that land use conflicts could potentially be more contentious given this scarcity. Centralized solar energy uses considerably larger land space than its decentralized counterparts. For example, the WEC (2016) states that in the U.S., utility-scale solar requires an average of $36\,m^2$ per MW of which 30 is directly used. Households' rooftop surfaces are suited for small-scaled solar energy systems so would minimize the opportunity costs involved with competing land uses.
- *Household solar installations do not benefit from economies of scale*: There are many types of housing which will affect the size of a given household solar energy system. But most residential installations will be individual projects, and this means that these smaller-scaled systems do not benefit from the economies of scale as larger projects would.

Mainstreaming and the multi-level perspective

'The mainstream' (used as a noun) is conventionally thought of as those elements of society that are shared by most people as a normal part of life (Lexico, 2019; Cambridge University Press, 2019a). In this train of thought, anything that is considered 'mainstream' (used as an adjective) or which has been 'mainstreamed' (used as a verb), can be thought of as being integrated into 'the mainstream'. Therefore, 'mainstreaming' is the process of making such aspects of society accepted by the majority as a normal part of life (Cambridge University Press, 2019b). So, applying this to solar energy means that adopting the technologies and using the resource become normalized by most people. But, how do these notions relate to the MLP?

Authors such as Seyfang and Smith (2007), Smith, Voß and Grin (2010), Coenen, Raven and Verbong (2010), Yanosek (2012), and Green, Skerlos and Winebrake (2014) state that there is a difference between niches and the mainstream; niches are embedded within the mainstream (Egmond, Jonjers & Kok, 2006a). As hinted earlier in the chapter, the regime is poised as 'the mainstream' in the MLP. Smith, Voß and Grin (2010), Shove and Walker (2010) and Coenen et al. (2010) even directly identify the regime as the mainstream, and Smith, Raven and Verbong and Lockwood et al. (2013) further describe it as the mainstream ways of fulfilling social functions.

The regime provides the 'selection environment' for innovations and competition selects the winning innovations (Markard & Truffer, 2008; Smith et al. 2010; Lockwood et al., 2013). Politics, culture, geography and the economy also influence the winning innovations (O'Connor, 2010). The creation and evolution of niches therefore combine 'variation' and 'selection' (Schot, 1998). This is seen in Figure 2.2 where the arrows begin to settle from their haphazard directional orientations in the niche-regime space.

There is no formal definition of mainstreaming in the MLP but an innovation's conceptual migration from its niche and run-up to 'breaking through' into the regime can be thought of as the process of mainstreaming, that is, Phases 1 to 3 in Figure 2.2. This means that the book's 'mainstreaming' in the title refers to household SWHs and PV applications evolving out of these niches. Figure 2.2's annotation below the right regime-polygon, together with its line of symmetry on the horizontal axis of time, also shows that mainstreaming is a process and not an event in time.

The diffusion of innovation theory developed by Everett Rogers is not applied in the book, but it is worth outlining how mainstreaming is linked to it, given the importance of adopting innovations in the MLP; work such as Rogers (1983), Richerson, Mulder and Vila (2001), Robinson (2009), and Mani and Dhingra (2012) give more details on the diffusion of innovations.

An adopting market is made up of groups segmented into the innovators, early adopters, early majority, late majority and laggards, and each has their own characteristics. They make up potentially 2.5%, 13.5%, 24%, 35% and 16% of the market respectively (Rogers, 1983). Based on this distribution, if

38 *Island energy transitions and cultures*

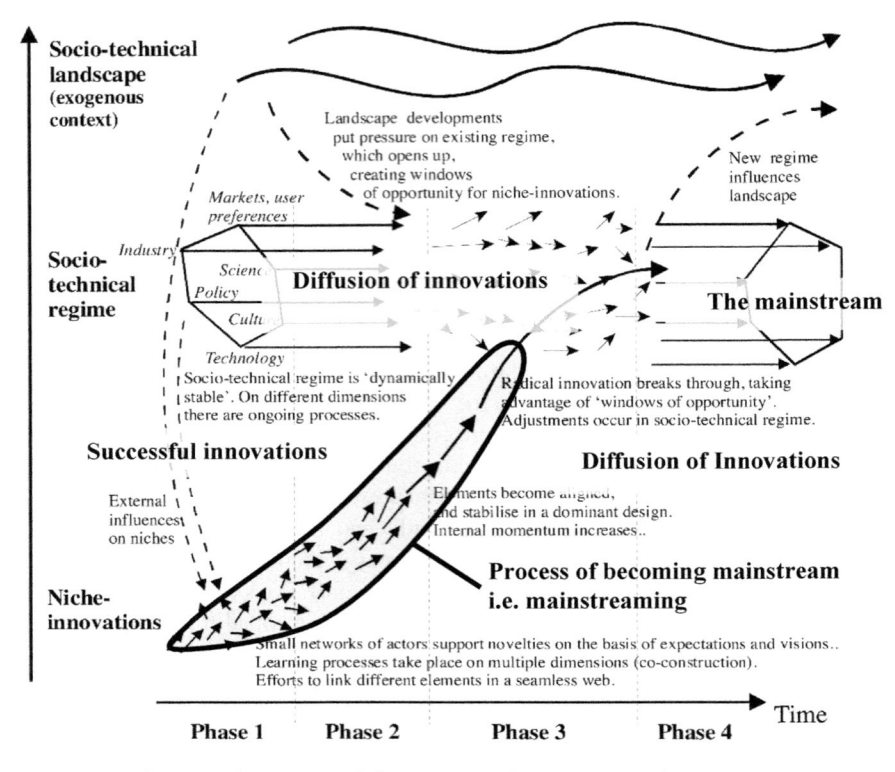

Figure 2.2 Showing the MLP and the conceptual positioning of mainstreaming.

an innovation is adopted by >2.5% the market; it enters the early adopters' segment; >16% into the early majority's; >40% into the late majority's; and >75% into the laggards'.

The early majority is thought to be the beginning of the mainstream market (Moore, 1991; Egmond, Jonkers & Kok 2006a; Goldenberg et al., 2006). Therefore, a theoretical mainstream threshold can be thought to exist at 16% adoption. A caveat however is the fact that an adoption market varies so this 16% threshold is not to be an absolute figure. Nevertheless, the diffusion of innovation work offers insight into what the mainstream market segments are like.

Mainstream adopters are less willing than the early adopters to adopt innovations; are very practical and solutions-oriented; will invest in an innovation that improves the existing services they are provided with; and if they must adopt an innovation, will usually seek out improvements that reduce the potential deviation from their norms (Egmond, Jonkers & Kok, 2006b).

Therefore, the mainstream can be referred to as 'average', widespread, involved in daily use and rooted in convenience (Green, Skerlos & Winebrake 2014) – as hinted from the more conventional definition presented in this section's first paragraph. Several of the mainstream's characteristics are also consistent with the incremental development of innovations and changes in regime-states highlighted earlier in the book's characterization of the MLP.

Figure 2.3 puts the concept of mainstreaming into the case studies' context by illustrating where the three case study islands would fit based on their respective adoption of SWHs and PV; Trinidad has only minimal household adoption, Barbados has mainstreamed SWHs and is now adopting PV, and Oʻahu has mainstreamed PV and is now adopting batteries.

Mainstreaming and an energy transition in the MLP are not the same conceptually. An energy transition is a change from one regime-orientation into another brought about by the landscape exerting change-pressures on the incumbent regime which opens spaces for niche-innovations to enter and subsequently modify this regime-state. Mainstreaming, however, specifically refers to the evolution of an innovation out of its niche and subsequent entry into the regime.

These theoretical ideas suggest that mainstreaming is conceptually founded on several assumptions inherited from the MLP:

- An energy transition is a change in regime-state.
- There are cultural, policy-based, technological, market-related, scientific and industrial elements that affect regimes and niches.
- An energy transition is made up of cultural, policy-based, technological, market-related, industrial and scientific transitions.

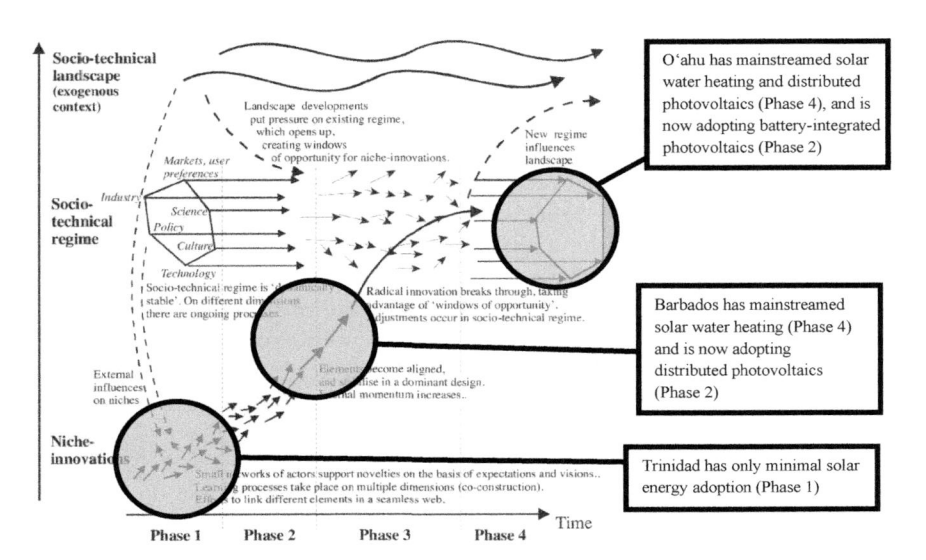

Figure 2.3 Showing the positioning of where the case study islands would fit in the MLP relative to their energy transitions through residential solar energy.

- A transition in one element will affect the others since they are co-evolutionary.
- Mainstreaming an innovation can catalyse a regime's cultural, policy-based, technological, market-oriented, industrial and/or scientific transitions. A caveat is that a technology could be mainstreamed without catalysing an energy transition.

Summary

'Energy transition' is a global buzzword that is more frequently being used in reference to the world's need to move away from fossil fuels to more renewable alternatives given the implications of anthropogenic environmental change and unsustainable energy use. Whilst the concept is very suggestive of a shift in some form in an energy sector, defining an energy transition is by no means a clear-cut task because of how complex energy systems are.

This complexity extends conceptually to interchangeable terminologies and jargon as well as sector(s) of interest, for example 'low-carbon transition' versus 'sustainable energy transition' and transitions in the power generation sector versus wider energy sector respectively.

One of the most common ways of approaching energy transitions is through the MLP. It is a theory/framework that comprises a socio-technical landscape, socio-technical regime and niche-innovations. It frames an energy transition as the product of the landscape exerting change-pressures on the regime to create opportunities for niche-innovations to enter the regime to catalyse the energy transition.

The socio-technical regime, however, is the structure that undergoes the shift which defines the energy transition. It is made up of the interaction between policies, cultures, technologies, industries, science and markets. So, an energy transition in the context of the MLP refers to a realignment of the incumbent regime's six elements. This regime plane is also 'the mainstream'. Therefore, innovations that evolve out of their niches and into the regime would have undergone the process of mainstreaming – and this is the theoretical basis for looking at the mainstreaming of household SWHs and PV in small, tropical islands for the book.

However, given that the regime is the mainstream structure of the MLP, and 'culture' is an explicit mainstream element, this warrants more elaboration considering the book's title. So, the next chapter builds on this chapter's characterization of 'the mainstream' by making the linkages between energy transitions, regimes, mainstreaming and 'culture'.

References

Andrews-Speed, P. (2016). Applying Institutional Theory to the Low-carbon Energy Transition. *Energy Research and Social Science*, *13*: 216–225. Retrieved from https://doi.org/10.1016/j.erss.2015.12.011.

Beerepoot, M. & Beerepoot, N. (2007). Government Regulation as an Impetus for Innovation: Evidence from Energy Performance Regulation in the Dutch Residential Building Sector. *Energy Policy*, *35*(10): 4812–4825. Retrieved from https://doi.org/10.1016/j.enpol.2007.04.015.

Berkhout, F., Smith, A. & Stirling, A. (2004). Socio-technical Regimes and Transition Contexts. In B. Elzen, F. W. Geels & K. Green (Eds.), *System Innovation and the Transition to Sustainability: Theory, Evidence and Policy* (pp. 48–75). Cheltenham: Edward Elgar. Retrieved from: https://doi.org/10.4337/9781845423421.

Brass, D. J. & Burkhardt, M. E. (1993). Potential Power and Power Use: An Investigation of Structure and Behaviour. *Academy of Management Journal*, *36*(3): 441–470. Retrieved from www.jstor.org/stable/256588 (DOI: 10.2307/256588).

Bridge, G., Bouzarovski, S., Bradshaw, M. & Eyre, N. (2013). Geographies of Energy Transition: Space, Place and the Low-carbon Economy. *Energy Policy*, *53*: 331–340. Retrieved from https://doi.org/10.1016/j.enpol.2012.10.066.

Brinkworth, M. (1985). *Tomorrow's World Energy*. London: The British Broadcasting Corporation.

Buchanan, M. (2017). Energy Transitions. *Nature Physics*, *13*: 1144. Retrieved from https://doi.org/10.1038/nphys4329.

Cambridge University Press (2019a). *Meaning of Mainstream in English*. Retrieved from https://dictionary.cambridge.org/dictionary/english/mainstream.

Cambridge University Press (2019b). *Meaning of Mainstreaming in English*. Retrieved from https://dictionary.cambridge.org/dictionary/english/mainstreaming.

Cherp, A., Vinichenko, V., Jewell, J., Brutschin E. & Sovacool, B (2018). Integrating Techno-economic, Socio-technical and Political Perspectives on National Energy Transitions: A Meta-theoretical Framework. *Energy Research & Social Science*, *37*: 175–190. Retrieved from https://doi.org/10.1016/j.erss.2017.09.015.

Coenen, L., Raven, R. & Verbong, G. (2010). Local Niche Experimentation in Energy Transitions: A Theoretical and Empirical Exploration of Proximity Advantages and Disadvantages. *Technology in Society*, *32*(4): 295–302. Retrieved from https://doi.org/10.1016/j.techsoc.2010.10.006.

Cressman, D. (2009). *A Brief Overview of Actor-network Theory: Punctualization, Heterogeneous Engineering & Translation*. Presentation published by ACT Lab/Centre for Policy Research on Science & Technology (CPROST), School of Communication, Simon Fraser University, British Columbia, Canada. Retrieved from https://summit.sfu.ca/item/13593.

Crosby A. W. (2006). *Children of the Sun: A History of Humanity's Unappeasable Appetite for Energy*. New York: W. W. Norton and Company Incorporated.

CSO (Central Statistical Office). (2012). *Trinidad and Tobago 2011 Population and Housing Census Demographic Report*. Port of Spain, Trinidad and Tobago: Ministry of Planning and Sustainable Development, Government of the Republic of Trinidad and Tobago. Retrieved from www.tt.undp.org/content/dam/trinidad_tobago/docs/DemocraticGovernance/Publications/TandT_Demographic_Report_2011.pdf.

Daghfous, N., Petrof, J. V. & Pons, F. (1999). Values and Adoption of Innovations: A Cross-cultural Study. *Journal of Consumer Marketing*, *16*(4): 314–331. Retrieved from https://doi.org/10.1108/07363769910277102.

Dirks, G., King, L., Laird, F., Lloyd, J., Lovering, J., Nordhaus, T., Pielke Jr., R., Román, M., Sarewitz, D., Shellenberger, M., Singh, K. & Trembath, A. (2014).

High-Energy Innovation: A Climate Pragmatism Project. Oakland, CA: The Breakthrough Institute. Retrieved from https://thebreakthrough.org/articles/high-energy-innovation.

Dumas, M., Rising, J. & Urpelainen, J. (2016). Political Competition and Renewable Energy Transitions Over Long Time Horizons: A Dynamic Approach. *Ecological Economics*, *124*: 175–184. Retrieved from https://doi.org/10.1016/j.ecolecon.2016.01.019.

Egmond, C., Jonkers, R. & Kok, G. (2006a). A Strategy and Protocol to Increase Diffusion of Energy-related Innovations into the Mainstream of Housing Associations. *Energy Policy*, *34*(18): 4042–4049. Retrieved from https://doi.org/10.1016/j.enpol.2005.10.001.

Egmond, C., Jonkers, R. & Kok, G. (2006b). One Size Fits All? Policy Instruments Should Fit the Segments of Target Groups. *Energy Policy*, *34*(18): 3464–3474. Retrieved from https://doi.org/10.1016/j.enpol.2005.07.017.

Eid, C., Guillén, J. R., Marín, P. F. & Hakvoort, R. (2014). The Economic Effect of Electricity Net-metering with Solar PV: Consequences for Network Cost Recovery, Cross Subsidies and Policy Objectives. *Energy Policy*, *75*: 244–254. Retrieved from https://doi.org/10.1016/j.enpol.2014.09.011.

Elias, R. J. & Victor, D. G. (2005). *Energy Transitions in Developing Countries: A Review of Concepts and Literature*. Working Paper #40, Program on Energy and Sustainable Development, Center for Environmental Science and Policy. Stanford, CA: Stanford University. Retrieved from https://pesd.fsi.stanford.edu/sites/default/files/energy_transitions.pdf.

Eni. (2016). *Energy Sources*. Eniscuola. Retrieved from www.eniscuola.net/wp-content/uploads/2011/02/pdf_energy_knowledge_11.pdf.

Espinasa, R. & Humpert, M. (2016). *Energy Dossier: Trinidad and Tobago*. Inter-American Development Bank. Retrieved from https://publications.iadb.org/en/energy-dossier-trinidad-and-tobago.

European Commission (2011). *World and European Energy and Environment Transition Outlook*. Brussels, Belgium. Retrieved from https://ec.europa.eu/research/social-sciences/pdf/policy_reviews/publication-weto-t_en.pdf.

Fouquet, R. (2011). *Lessons from History for Transitions to a Low Carbon Economy*. Public Policy Briefings 2011–04, Basque Centre for Climate Change, Leioa, Spain. Retrieved from www.bc3research.org/policybriefings/2011-02.html.

Fouquet, R. (2016). Historical Energy Transitions: Speed, Prices and System Transformation. *Energy Research and Social Science*, *22*: 7–12. Retrieved from https://doi.org/10.1016/j.erss.2016.08.014.

Fouquet, R. & Pearson, P. J. G. (2012). Past and Prospective Energy Transitions: Insights from History. *Energy Policy*, *50*: 1–7. Retrieved from https://doi.org/10.1016/j.enpol.2012.08.014.

Fuchs, G. & Hinderer, N. (2016). One Or Many Transitions: Local Electricity Experiments in Germany. *Innovation: The European Journal of Social Science Research*, *29*(3): 320–336. Retrieved from https://doi.org/10.1080/13511610.2016.1188683.

Gates, B. (2015). *Energy Innovation: Why We Need It and How to Get It*. The Gates Notes. Retrieved from www.gatesnotes.com/~/media/Files/Energy/Energy_Innovation_Nov_30_2015.pdf?la=en.

Geels, F. W. (2002). Technological Transitions as Evolutionary Reconfiguration Process: A Multi-level Perspective and a Case Study. *Research Policy*, *31*(8–9): 1257–1274. Retrieved from https://doi.org/10.1016/S0048-7333(02)00062-8.

Geels, F. W. (2004). From Sectoral Systems of Innovation to Socio-technical Systems: Insights about Dynamics and Change from Sociology and Institutional Theory. *Research Policy*, *33*(6–7): 897–920. Retrieved https://doi.org/10.1016/j.respol.2004.01.015.

Geels, F. W. (2006). The Hygienic Transition from Cesspools to Sewer Systems (1840–1930): The Dynamics of Regime Transformation. *Research Policy*, *35*(7): 1069–1082. Retrieved from https://doi.org/10.1016/j.respol.2006.06.001.

Geels, F. W. (2018). Disruption and Low-carbon System Transformation: Progress and New Challenges in Socio-technical Transitions Research and the Multi-Level Perspective. *Energy Research & Social Science*, *37*: 224–231. Retrieved from https://doi.org/10.1016/j.erss.2017.10.010.

Geels, F. W. & Schot, J. (2007). Typology of Socio-technical Transition Pathways. *Research Policy*, *36*(3): 399–417. Retrieved from https://doi.org/10.1016/j.respol.2007.01.003.

Georges, E., Braun, J. E., Groll, E., Horton, W. T. & Lemort, V. (2014). *Impact of Net Metering Programs on Optimal Load Management in US Residential Housing – A Case Study*. Paper presented at the 9th International Conference on System Simulation in Buildings, Liège, Belgium. Retrieved from http://orbi.ulg.ac.be/bitstream/2268/178419/1/P64v2.pdf.

Gismondi, M. (2018). Historicizing Transitions: The Value of Historical Theory to Energy Transition Research. *Energy Research & Social Science*, *38*: 193–198. Retrieved from https://doi.org/10.1016/j.erss.2018.02.008.

Goldenberg, J., Libai, B., Muller, E. & Peres, R. (2006). Blazing Saddles: The Early and Mainstream Markets in the High-tech Product Life Cycle. *Israel Economic Review*, *4*(2): 85–108. Retrieved from https://ssrn.com/abstract=2178985.

Grayson, M. (2017). Energy Transitions: On a Global Scale, Reducing Fossil-fuel Dependence is More than Just a Technological Challenge. *Nature*, *551*: S133. Retrieved from www.nature.com/articles/d41586-017-07507-y (DOI: 10.1038/d41586-017-07507-y).

Green, E. H., Skerlos, S. J. & Winebrake, J. J. (2014). Increasing Electric Vehicle Policy Efficiency and Effectiveness by Reducing Mainstream Market Bias. *Energy Policy*, *65*: 562–566. Retrieved from https://doi.org/10.1016/j.enpol.2013.10.024.

Greenhalgh, T. & Stones, R. (2010). Theorising Big IT Programmes in Healthcare: Strong Structuration Theory Meets Actor-network Theory. *Social Science and Medicine*, *70*(9): 1285–1294. Retrieved from https://doi.org/10.1016/j.socscimed.2009.12.034.

Grubler, A. (2012a). Energy Transitions Research: Insights and Cautionary Tales. *Energy Policy*, *50*: 8–16. Retrieved from https://doi.org/10.1016/j.enpol.2012.02.070.

Grubler, A. (2012b). Grand Designs: Historical Patterns and Future Scenarios of Energy Technological Change. Historical Case Studies of Energy Technology Innovation in Chapter 24. In A. Grubler, F. Aguayo, K. S. Gallagher, M. Hekkert, K. Jiang, L. Mytelka, L. Neij, G. Nemet & C. Wilson (Eds.), *Global Energy Assessment – Toward a Sustainable Future* (pp. 1665–1744). Cambridge, UK: Cambridge University Press and Laxenburg, Austria: The International Institute for Applied Systems Analysis. Retrieved from www.iiasa.ac.at/web/home/research/researchPrograms/TransitionstoNewTechnologies/01_Grubler_Transitions__WEB.pdf.

Hargreaves T., Longhurst, N. & Seyfang, G. (2013). Up, Down, Round and Round: Connecting Regimes and Practices in Innovation for Sustainability. *Environment and Planning A, 45*: 402–420. Retrieved from https://doi.org/10.1068%2Fa45124.

Hauff, J., Bode, A., Neumann, D. & Haslauer, F. (2014). *Global Energy Transitions: A Comparative Analysis of Key Countries and Implications for the International Energy Debate*. Berlin, Germany: World Energy Council and ATKearney. Retrieved from www.atkearney.com/documents/10192/5293225/Global+Energy+Transitions.pdf/220e6818-3a0a-4baa-af32-8bfbb64f4a6b.

Herrera, H. (2015). *Sustainable Energy Transition and Climate Change Vulnerabilities: A Resilience Perspective*. Paper presented at the Systems Dynamics Society Conference, Delft, The Netherlands. Retrieved from www.systemdynamics.org/assets/conferences/2016/proceed/papers/P1107.pdf.

Hiemstra-van der Horst, G. & Hovorka, J. (2008). Reassessing the 'Energy Ladder': Household Energy Use in Maun, Botswana. *Energy Policy, 36*(9): 3333–3344. Retrieved from https://doi.org/10.1016/j.enpol.2008.05.006.

Horta, A., Wilhite, H., Schmidt, L. & Baritaux, F. (2014). Socio-technical and Cultural Approaches to Energy Consumption. *An Introduction: Nature and Culture, 9*(2): 115–121. Retrieved from https://doi.org/10.3167/nc.2014.090201.

IEA and IRENA (International Energy Agency and International Renewable Energy Agency). (2017). *Perspectives for the Energy Transition: Investment Needs for a Low-Carbon Energy System*. German Federal Ministry for Economic Affairs and Energy, Berlin, Germany. Retrieved from www.irena.org/-/media/Files/IRENA/Agency/Publication/2017/Mar/Perspectives_for_the_Energy_Transition_2017.pdf?la=en&hash=56436956B74DBD22A9C6309ED76E3924A879D0C7.

Ince, D. (2018). *Barbados National Energy Policy (2017–2037)*. Presented to the Division of Energy and Telecommunications, Prime Minister's Office, Government of Barbados, Barbados. Retrieved from www.energy.gov.bb/web/component/docman/doc_download/86-barbados-national-energy-policy-2017-2037.

Jenkins, K., Sovacool, B. K. & McCauley, D. (2018). Humanizing Socio-technical Transitions through Energy Justice: An Ethical Framework for Global Transformative Change. *Energy Policy, 117*: 66–74. Retrieved from https://doi.org/10.1016/j.enpol.2018.02.036.

Kemp, R. (1994). Technology and the Transition to Environmental Sustainability: The Problem of Technological Regime Shifts. *Futures, 26*(10): 1023–1046. Retrieved from https://doi.org/10.1016/0016-3287(94)90071-X.

Kern, F. & Smith, A. (2008). Restructuring Energy Systems for Sustainability? Energy Transition Policy in the Netherlands. *Energy Policy, 36*(11): 4093–4103. Retrieved from https://doi.org/10.1016/j.enpol.2008.06.018.

Klitkou, A., Bolwig, S., Hansen, T. & Wessberg, N. (2015). The Role of Lock-in mechanisms in Transition Processes: The Case of Energy for Road Transport. *Environmental Innovations and Societal Transitions, 16*: 22–37. Retrieved from https://doi.org/10.1016/j.eist.2015.07.005.

Kuzemko, C., Lockwood, M., Mitchell, C. & Hoggett, R. (2016). Governing for Sustainable Energy System Change: Politics, Contexts and Contingency. *Energy Research & Social Science, 12*: 96–105. Retrieved from https://doi.org/10.1016/j.erss.2015.12.022.

Lacey, S. (2010, April 22). Why the Energy Transition is Longer Than We Admit. *Renewable Energy World.* Retrieved from www.renewableenergyworld.com/ugc/articles/2010/04/why-the-energy-transition-is-longer-than-we-admit.html.

Lawhon, M., & Murphy, J. T. (2011). Socio-technical Regimes and Sustainability Transitions: Insights from Political Ecology. *Progress in Human Geography, 36*(3): 354–378. Retrieved from https://doi.org/10.1177%2F0309132511427960.

Lexico. (2019). *Definition of Mainstream in English.* British & World English, Lexico.com. Retrieved from www.lexico.com/en/definition/mainstream.

Lockwood, M., Kuzemko, C., Mitchell, C. & Hoggett, R. (2013). *Theorising Governance and Innovation in Sustainable Energy Transitions.* Working Paper No. 1304. Energy Policy Group, University of Exeter, Exeter, United Kingdom. Retrieved from http://projects.exeter.ac.uk/igov/wp-content/uploads/2013/07/WP4-IGov-theroy-of-change.pdf.

Love, T. & Garwood, A. (2013). Electrifying Transitions: Power and Culture in Rural Cajamarca, Peru. In S. Strauss, S. Rupp & T. Love (Eds.), *Cultures of Energy: Power, Practices, Technologies* (1st ed.) (pp. 147–176). Walnut Creek, CA: Left Coast Press. Retrieved from https://doi.org/10.4324/9781315430850.

Mani, S. & Dhingra, T. (2012). Diffusion of Innovation Model of Consumer Behaviour: Ideas to Accelerate Adoption of Renewable Energy Sources by Consumer Communities in India. *Renewable Energy, 39*(1): 162–165. Retrieved from https://doi.org/10.1016/j.renene.2011.07.036.

Manning, R. A. (2015). *Renewable Energy's Coming of Age: A Disruptive Technology?* Washington, DC: Brent Scowcroft Center on International Security and Global Energy Center, Atlantic Council. Retrieved from www.files.ethz.ch/isn/195316/Renewable_Energy.pdf.

Markard, L. & Truffer, B. (2008). Technological Innovation Systems and the Multi-level Perspective: Towards an Integrated Framework. *Research Policy, 37*(4): 596–615. Retrieved from https://doi.org/10.1016/j.respol.2008.01.004.

Mattes, J., Huber, A. & Koehrsen, J. (2015). Energy Transitions in Small-scale Regions: What We Can Learn from a Regional Innovation Systems Perspective. *Energy Policy, 78*: 255–264. Retrieved from https://doi.org/10.1016/j.enpol.2014.12.011.

McVeigh, J. C. (1984). *Energy Around the World: An Introduction to Energy Studies: Global Resources, Needs, Utilization.* Oxford, UK: Pergamon Press Limited.

Miller, C. A., Iles, A. & Jones, C. F. (2013). The Social Dimensions of Energy Transitions. *Science as Culture, 22*(2): 135–148. Retrieved from https://doi.org/10.1080/09505431.2013.786989.

Moore, G. A. (1991). *Crossing the Chasm: Marketing and Selling High-Tech Products to Mainstream Customers* (Revised ed). New York: HarperCollins Publishers. Retrieved from http://soloway.pbworks.com/w/file/fetch/46715502/Crossing-The-Chasm.pdf.

Namias, O. (2008). The Hidden Dimensions of Electrical Architecture. In M. Rüdiger (Ed.), *The Culture of Energy* (pp. 93–102). Newcastle: Cambridge Scholars Publishing.

Nimmo, R. (2011). Actor-network Theory and Methodology: Social Research in a More-than-human World. *Methodological Innovations Online, 6*(3): 108–119. Retrieved from: https://doi.org/10.4256%2Fmio.2011.010.

O'Connor, P. A. (2010). *Energy Transitions*. The Pardee Papers, No. 12. Boston, MA: The Frederick S. Pardee Center for the Study of the Longer-Range Future, Boston University. Retrieved from www.bu.edu/pardee/pardee-paper-012-energy/.

O'Connor, P. A. & Cleveland, C. J. (2014). U.S. Energy Transitions 1780–2010. *Energies*, 7: 7955–7993. Retrieved from https://doi.org/10.3390/en7127955.

Pain, S. (2017). Power Through the Ages: From Elbow Grease to Electricity, Horsepower to Hydropower, Energy Has Played a Crucial Part in Shaping Society. *Nature*, 551: S134-S137. Retrieved from www.nature.com/articles/d41586-017-07506-z (DOI: 10.1038/d41586–017–07506-z).

Palm, J. (2006). Development of Sustainable Energy Systems in Swedish Municipalities: A Matter of Path Dependency and Power Relations. *Local Environment: The International Journal of Justice and Sustainability*, 11(4): 445–457. Retrieved from https://doi.org/10.1080/13549830600785613.

Pérez-Lombard, L., Ortiz, J. & Pout, C. (2008). A Review on Building Energy Consumption Information. *Energy and Buildings*, 40(3): 394–398. Retrieved from https://doi.org/10.1016/j.enbuild.2007.03.007.

Pollack, J., Costello, K. & Sankaran, S. (2013). Applying Actor-Network Theory as a Sensemaking Framework for Complex Organisational Change Programs. *International Journal of Project Management*, 31(8): 1118–1128. Retrieved from https://doi.org/10.1016/j.ijproman.2012.12.007.

Raven, R., Schot, J. & Berkhout, F. (2012). Space and Scale in Socio-technical Energy Transitions. *Environmental Innovations and Societal Transitions*, 4: 63–78. Retrieved from https://doi.org/10.1016/j.eist.2012.08.001.

READ (Research and Economic Analysis Division). (2019). *Hawai'i Energy Statistics*. The Energy Dashboard. Department of Business, Economic Development and Tourism, Hawai'i State Energy Office, State of Hawai'i, Honolulu. Retrieved from https://energy.Hawai'i.gov/resources/dashboard-statistics.

REN21 (Renewable Energy Policy Network for the 21st Century). (2017). *Advancing the Global Renewable Energy Transition: Highlights of the REN21 Renewables 2017 Global Status Report in Perspective*. Paris, France: REN21 Secretariat, United Nations Environment. Retrieved from www.ren21.net/wp-content/uploads/2019/05/GSR2017_Highlights_English.pdf.

Rhodes, R. (2007). *Energy Transitions: A Curious History*. Presentation published by the Center for International Security and Cooperation, Stanford University and given at The Security Implications of Increased Global Reliance on Nuclear Power Conference. Stanford, CA: Stanford University. Retrieved from https://cisac.fsi.stanford.edu/publications/energy_transitions_a_curious_history.

Richerson, P. J., Mulder, M. B. & B. Vila, B. (2001). *Diffusion of Innovations*. In *Principles of Human Ecology* (pp. 350–364). California: University of California, Davis. Retrieved from www.des.ucdavis.edu/faculty/Richerson/BooksOnline/101text.htm.

Robinson, L. (2009). *A Summary of Diffusion of Innovations*. Retrieved from www.enablingchange.com.au/Summary_Diffusion_Theory.pdf.

Rogers, E. M. (1983). The Innovation-Decision Process. In *Diffusion of Innovations* (3rd ed.) (pp. 163–209). New York: The Free Press.

Sarrica, M., Brondi, S., Cottone P. & Mazzara, B. M. (2016). One, No One, One Hundred Thousand Energy Transitions in Europe: The Quest for a Cultural

Approach. *Energy Research and Social Science, 13*: 1–14. Retrieved from https://doi.org/10.1016/j.erss.2015.12.019.

Schot, J. W. (1998). The Usefulness of Evolutionary Models for Explaining Innovation: The Case of The Netherlands in the Nineteenth Century. *History of Technology, 14*(3): 173–200. Retrieved from https://doi.org/10.1080/07341519808581928.

Schot, J. W., Hoogma, R. & Elzen, B. (1994). Strategies for Shifting Technological Systems: The Case of the Automobile System. *Futures, 26*(10): 1060–1076. Retrieved from https://doi.org/10.1016/0016-3287(94)90073-6.

Seyfang, G. & Smith, A. (2007). Grassroots Innovations for Sustainable Development: Towards a New Research and Policy Agenda. *Environmental Politics, 16*(4): 584–603. Retrieved from https://doi.org/10.1080/09644010701419121.

Shove, E. & Walker, G. (2010). Governing Transitions in the Sustainability of Everyday Life. *Research Policy, 39*(4): 471–476. Retrieved from https://doi.org/10.1016/j.respol.2010.01.019.

Smil, V. (2016). Examining Energy Transitions: A Dozen Insights Based on Performance. *Energy Research and Social Science, 22*: 194–197. Retrieved from https://doi.org/10.1016/j.erss.2016.08.017.

Smith, A., Stirling, A. & Berkhout, F. (2005). The Governance of Sustainable Sociotechnical Transitions. *Research Policy, 34*(10): 1491–1510. Retrieved from https://doi.org/10.1016/j.respol.2005.07.005.

Smith, A., Voß, J. & Grin, J. (2010). Innovation Studies and Sustainability Transitions: The Allure of the Multi-level Perspective and its Challenges. *Research Policy, 39*(4): 435–448. Retrieved from https://doi.org/10.1016/j.respol.2010.01.023.

Sovacool, B. K. (2016). How Long Will It Take? Conceptualizing the Temporal Dynamics of Energy Transitions. *Energy Research & Social Science, 13*: 202–215. Retrieved from https://doi.org/10.1016/j.erss.2015.12.020.

Sovacool, B. K. & Geels, F. W. (2016). Further Reflections on the Temporality of Energy Transitions: A Response to Critics. *Energy Research & Social Science, 22*: 232–237. Retrieved from https://doi.org/10.1016/j.erss.2016.08.013.

Sovacool, B. K. & Hess, D. J. (2017). Ordering Theories: Typologies and Conceptual Frameworks for Socio-technical Change. *Social Studies of Sciences, 47*(5): 703–750. Retrieved from https://doi.org/10.1177%2F0306312717709363.

Steg, L., Perlaviciute, G. & van der Werff, E. (2015). Understanding the Human Dimensions of a Sustainable Energy Transition. *Frontiers in Psychology, 8*(Article 805): 1–17. Retrieved from https://doi.org/10.3389/fpsyg.2015.00805.

Strebel, F. (2011). Inter-governmental Institutions as Promoters of Energy Policy Diffusion in a Federal Setting. *Energy Policy, 39*(1): 467–476. Retrieved from https://doi.org/10.1016/j.enpol.2010.10.028.

Truffer, B. & Coenen, L. (2012). Environmental Innovation and Sustainability Transitions in Regional Studies. *Regional Studies, 46*(1): 1–21. Retrieved from https://doi.org/10.1080/00343404.2012.646164.

Truffer, B., Murphy J. T. & Raven, R. (2015). The Geography of Sustainability Transitions: Contours of an Emerging Theme. *Environmental Innovations and Societal Transitions, 17*: 63–72. Retrieved from https://doi.org/10.1016/j.eist.2015.07.004.

Urry, J. (2014). The Problem of Energy. *Theory, Culture and Society, 31*(5): 3–20. Retrieved from https://doi.org/10.1177%2F0263276414536747.

WEC (World Energy Council). (2016). *World Energy Resources- Solar 2016* (pp. 26–27). London. Retrieved from www.worldenergy.org/assets/images/imported/2016/10/World-Energy-Resources-Full-report-2016.10.03.pdf.

WEF (World Economic Forum). (2013). *Energy Vision 2013. Energy Transitions: Past and Future*. Geneva, Switzerland. Retrieved from www3.weforum.org/docs/WEF_EN_EnergyVision_Report_2013.pdf.

WEF (World Economic Forum). (2018). *Fostering Effective Energy Transition: A Fact-Based Framework to Support Decision-Making* (Insight Report). Geneva, Switzerland. Retrieved from www3.weforum.org/docs/WEF_Fostering_Effective_Energy_Transition_report_2018.pdf.

Westphal, K. (2012). *Globalising the German Energy Transition*. SWP Comment 2012/C 40. German Institute for International and Security Affairs, Berlin, Germany. Retrieved from www.swp-berlin.org/fileadmin/contents/products/comments/2012C40_wep.pdf.

Winskel, M., Radcliffe, J., Skea, J. & Wang, X. (2014). Remaking the UK's Energy Technology Innovation System: From the Margins to the Mainstream. *Energy Policy*, 68: 591–602. Retrieved from https://doi.org/10.1016/j.enpol.2014.01.009.

Yanosek, K. (2012). Policies for Financing the Energy Transition. *Daedalus, the Journal of the American Academy of Arts and Sciences*, 141(2): 94–104. Retrieved from https://doi.org/10.1162/DAED_a_00149.

Zailani, S., Govindan, K., Iranmanesh, M., Shaharudin, M. R. & Chong, Y. S. (2015). Green Innovation Adoption in Automotive Supply Chain: The Malaysian Case. *Journal of Cleaner Production*, 108(Part A): 1115–1122. Retrieved from https://doi.org/10.1016/j.jclepro.2015.06.039.

3 Energy transitions and mainstream energy cultures

Introduction

Festivals, clothing, artwork and music are all popularly and conventionally thought of as representations of 'culture', for example the costumes worn by the individuals in Figure 3.1. They are indeed cultural, but they are expressions of cultures. So, what is 'culture' really?

Like the concept of an 'energy transition' explored in Chapter 2, there is no universal definition for 'culture' (Tharp, 2012; Arnesen, 2013). As a standalone concept, it is difficult to understand and engage with scientifically because it is not directly observable (Dubois, 1972; Gallivan & Srite, 2005), is a human phenomenon that most are unaware of (Arnesen, 2013), and everyday life in its broadest sense is difficult to research (Strengers, 2010).

'Culture' has both invisible (e.g. values, attitudes, assumption, beliefs) and visible (e.g. artefacts, behaviours) elements (Tharp, 2012). It is about organization (White, 1943), that is, organized norms, values, knowledge, beliefs, art, morals, customs, material items, as well as any other aspects of expression that are acquired by being a member of society (Zion & Kozleski, 2005; Feng & Mu, 2010; Broesch & Hadley, 2012).

To illustrate the complexity of such socially acquired traits, think of cultures as phenomena founded on values and norms (Gallivan & Srite, 2005) – where norms are values that are socially expressed through routinely accepted social behavioural patterns (Dubois, 1972) and values are individuals' permanent beliefs (Daghfous, Petrof & Pons, 1999). A culture, then, can be viewed as a system of values that is shared people (Dubois, 1972); there are intrinsic and extrinsic values where the latter comes from other people (Chatterton, 2011). So, an individual's values are linked to a wider system of social values (Daghfous et al., 1999).

Further, values are the underlying agents of behaviours (Spencer-Oatey, 2012) and Sekiguchi and Nakamaru (2011) describe behaviours as observable actions. In the context of energy-related behaviours, Stephenson et al. (2010) suggest that they are commonly displayed through the adoption of technologies, for example the SWHs and PV being looked at in this book.

Figure 3.1 Showing an example of the conventional expressions of 'culture'.

But attitudes are the internal and unobservable reflections of these actions/ behaviours (Sekiguchi & Nakamaru, 2011) – and whilst there is a relationship between attitudes and behaviours (Berkowitz & Haines, 1984), there is also an inconsistency between both (Leonard-Barton, 1981). Sekiguchi and Nakamaru (2011) provide an example where consumers can adopt the socially acceptable behavioural norm regardless of their attitude and this leads to the inconsistency.

Berkowitz and Haines (1984) describe work that frame behaviours as being driven by intentions which are in turn affected by attitudes; attitudes by beliefs; and beliefs by cognitive responses. Klockner (2013) further contextualizes a belief in this respect as the societal expectation that a given behaviour will result in a certain outcome whilst attitudes relate to the favourability of behavioural alternatives – with each having their respective outcomes.

It is these series of intricate perspectives that are encapsulated in the broad notion of what 'culture' is conventionally thought to be, that is, 'a way of life'. Nevertheless, cultural theorists have suggested that thoughts, actions and material products are its main elements (Tharp, 2012).

A culture is shared by at least two people (Spencer-Oatey, 2012; Neuliep, 2015) and any attitude, behaviour or activity that an individual develops, whether it is through instruction or imitation, is cultural. In that regard, 'culture' is learned through socialization as well as personal experiences but is not genetically inherited (Tharp, 2012; Spencer-Oatey, 2012).

Individuals are born into cultures, but they have the capacity to choose to belong to them as well (Zion & Kozleski, 2005). Geels and Schot (2007) acknowledge this cognitive independence but outline that social interactions tend to converge on a shared outlook, perception or socio-cognitive impression that becomes the dominant and socially accepted norm. An individual is therefore part of a culture if they hold the same values and behaviours that can be considered the norm (Spencer-Oatey, 2012).

The last paragraph's point on convergence hints at the cultural model of a mainstream culture. Cultural models are formed when cultural information groups around a specific domain (Broesch & Hadley, 2012). So, domains help individuals understand and organize their social world (recalling that 'culture' is an organization phenomenon) (Fryberg & Rhys, 2009).

Now, how do these thoughts begin to relate to energy and 'the mainstream'? The short answer is that the mainstream culture would be the one in which the majority's shared traits converge in cultural domains built on/in the historical energy transitions outlined in Chapter 2 that have led to the electricity systems of today. This is because energy production and consumption has set the context for society's relations with its human and non-human elements (Horta et al., 2014) and the introduction of electricity as well as its presence and usage are culture-specific (Winther, 2013).

Therefore, it is reasonable to suggest that people's interactions with their electricity supply, consumption and payment systems have developed cultures since people's habits would have been locked-in through their interactions within the culture, over history and with their material environments (Wallenborn & Wilhite, 2014).

'Mainstream culture' and the multi-level perspective

Before picking up on where the last section was heading with respect to 'mainstream culture', it is worth putting 'culture' into the context of the insularity of small, tropical islands. Spencer-Oatey (2012) and Neuliep (2015) outline that many think of 'culture' in terms of its geography, but it is affected by geography not defined by it.

However, Neuliep (2015) uses the term 'population' in relation to this relationship between cultures and geography; 'population' and 'culture' are not synonymous since 'population' implies some degree of geographic connectivity between its members. Therefore, in the context of small, tropical islands, the population is very much defined by the land-marine border though not necessarily the culture(s).

What this boundary does, however, is increase the frequency of interactions between members of the population such that the population has shared traits which, based on the thoughts described earlier, form the culture of the population residing on the island. These sorts of interactions create rather homogenous and close-knit collective identities in islands (King, 2009) because geographic proximity is multi-dimensional enough to reinforce them (Raven, Schot & Berkhout, 2012).

The fact that 'culture' can conceptually extend beyond islands' borders is not being disputed. This understanding is acknowledged and supported since it cannot be limited to one spatial scale of study. But using small-island insularity to conceptually define 'culture' is done on the basis that the culture of interest is that shared by the on-island population and the spatial fluidity of 'culture' as a concept has been recognized.

The dominant and most widespread assemblage of traits held by the resident population would be the mainstream culture of the island. But how is 'mainstream culture' related to the MLP used to approach energy transitions? It is firstly important to understand that 'culture' is not static; it continuously evolves and exhibits an adaptive capacity that includes resistance to changes (Zion & Kozleski, 2005; Heatwole, 2006; Tharp, 2012; Spencer-Oatey, 2012; Neuliep, 2015). Interestingly, this is quite like the nature of the socio-technical regime from Chapter 2.

In that regard, as an innovation enters the regime by being more widely adopted (Berkhout, Smith & Stirling, 2004), that is, becoming mainstreamed, social systems (and their culture(s)) need time to adjust their structures once they have begun assimilating the innovation since this reorientation allows them to retain their stability (Rogers et al., 2005).

So, based on the earlier description of what 'culture' is in its broadest sense, it is reasonable to suggest that 'culture' exists as a part of the socio-technical regime which is the MLP's mainstream structure – and in the case of this book, specifically defined as the residential electricity regime. This is explicitly supported by the fact that Figure 2.1 includes 'culture' as one of the six dimensions of the socio-technical regime.

However, landscape-forces, which are rather removed from the regime, are a diverse range of factors which include cultural norms (Geels & Schot, 2007; Lockwood et al. 2013). Foxon, Hammond and Pearson (2010) and Smith, Voß and Grin (2010) are examples of work which place 'culture' in the landscape of the MLP, and this is likely because of its structuration and slow rates of change (Loorbach, 2010).

So, relative to the planes of the MLP detailed in Chapter 2, that is, landscape, regime and niche-innovations, this section shows that 'culture' is complex because human beings are complex (Chatterton, 2011) and can exist in the regime as part of the mainstream, though it is arguably not as 'simple' enough to be solely a regime-level construct so also exists in the landscape. This also means that 'culture' is also not a completely exogenous element because it exists in the regime and not just in the landscape.

The narrative presented on 'culture' thus far has acknowledged its existence in energy transitions and that it spans the regime and landscape planes of the MLP. However, it is further expected that there are also niche-level cultures (as will be shown later in the chapter). Therefore, it is reasonable to assert that one reason why the role(s) of 'culture' in energy transitions was not clear in the literature reviewed is because it will co-exist in the MLP's planes in different capacities.

These considerations help with exploring the links between the broad notions of 'culture' and 'energy' as potentially linked concepts. As a result, defining 'mainstream culture' in relation to 'energy' required a secondary framework/theory and this is where the ECF comes in.

'Mainstream culture' and the energy cultures framework

Introducing the energy cultures framework

As outlined before, behaviours are central features of 'culture'. Behaviours involve acting based on beliefs and/or perceptions using, more often than not, tangible items – which has implications for the persons exhibiting the behaviour(s) as well as the tangible items used.

Arnesen (2013) states that exploring 'culture' cannot be limited to a single theory or discipline and Stephenson et al. (2010) acknowledge that microeconomics, behavioural economics, technology adoption, social and environmental psychology, amongst other sociological theories, are all relevant; it is because of this multiplicity of approaches that the understanding of energy and behaviour has grown over the decades (Lutzenhiser, 2008).

Behaviours can be thought of as manifestations of energy (White, 1943). Therefore, studying energy and behaviours should mean exploring how energy is used and what this energy use means to consumers (Lutzenhiser, 1992). Of the literature reviewed (e.g. Wallenborn, 2008; Arnesen, 2013; Strauss, Rupp & Love, 2013; and Sheller, 2014), Stephenson et al.'s (2010, 2015) ECF was the most detailed application of 'energy' and 'culture' as linked concepts.

The aim of the ECF is to give structure to the understanding of the drivers of energy-related behaviour and potential behaviour change interventions (Stephenson et al., 2010; Mirosa et al., 2011; Barton et al., 2013; Walton et al., 2014) – by exploring how people and their energy behaviours are embedded in daily life (Ford, Karlin & Frantz, 2016). Therefore, the ECF looks at 'energy behaviours' (Stephenson et al., 2015).

It does so by approaching them as energy cultures, that is, an assemblage of norms, material culture, practices and their external influences (see Figure 3.2). It was, however, more broadly renamed the Sustainability Cultures Framework given the model's applications to work beyond energy research (Stephenson, 2018). But, in the interest of consistency with the book's narrative and given the timing of when the body of research that informs the book was done, it will be referred to as the ECF.

The main alternative to using the ECF is the SPT, and it is commonly compared to the ECF (Stephenson, 2018). Works such as Shove (2010) as well as Shove and Walker (2010, 2014) touch on the SPT and its related

54 *Island energy transitions and cultures*

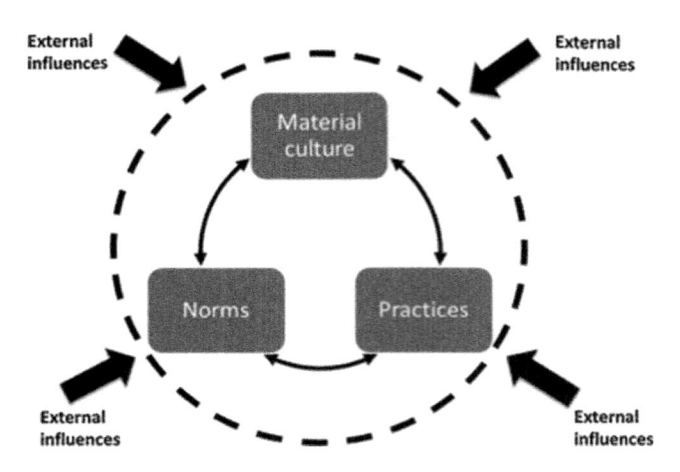

Figure 3.2 Showing a conceptualization of the ECF.

themes. The SPT looks at competencies, materials, connections and meanings (Sovacool & Hess, 2017). It does not look at behaviour but rather the practices rooted in daily activities (Shove & Walker, 2010; Ford et al., 2017) and investigates the co-structuration between human action and social structure relative to routinized practices (Sovacool & Hess, 2017; Stephenson, 2018).

The SPT focuses on a given practice and its social meaning (Shove & Walker, 2010) whilst the ECF approaches practices as a part of a larger context which is more in line with the assemblage of features that would be part of a regime and the mainstream in the context of the MLP. The ECF's consideration for the influence of an external contextual environment is another distinguishing element. According to Stephenson (2018), the ECF and SPT also differ because the latter does not consider non-routinized practices; focuses on cultural replication versus change; and is preoccupied with practices versus actors' cultural formations.

In terms of combining the MLP with a theory like SPT, Hargreaves, Longhurst and Seyfang (2013) outline that the MLP is a vertically structured theory but the SPT is horizontally oriented – and, as shown earlier, 'culture' transcends the planes of the MLP so is vertically structured as well. Additionally, the SPT treats actors as carriers of practices whilst the ECF is interested in the actors themselves and their agency – which complements the MLP since it focuses on the roles of actors in structural change (Ford et al., 2017).

Stephenson and colleagues think that a behavioural model should be versatile enough to have multi-scalar applications (Stephenson et al., 2015) because 'culture' is multi-scalar (Spencer-Oatey, 2012). The ECF is therefore useful because it is applicable to multiple analytical scales (Hopkins & Stephenson, 2014) – and this has been displayed through its empirical applications subsequent to the model's conceptualization in 2009 (Stephenson et al., 2015; see Mirosa et al., 2011; Barton et al., 2013; King et al., 2014 for examples).

The main assumption in the ECF is that a change in either norms, material culture or practices will stimulate a change in one or more of the other elements. But in terms of its limitations, there are several key points worth flagging.

The model considers the effect of external influences on its core of norms, material culture and practices. But the processes by which this core can influence the outlying external influences is not as clear. Also, the ECF does not explicitly consider time but instead captures snapshots of energy cultures. This is a major limitation since its norms, practices, material culture and their external influences are products of time-sensitive processes. Related to this is the lack of consideration for the actual processes that manifest norms, practices, material culture and external influences, for example cultural erasure, assimilation, retention or transmission.

Additionally, it does not address any differences in the strengths of expression of, and association between features that can be considered norms, material culture, practices and external influences. In that same breath, the conceptual boundaries between norms, material culture, practices and external influences are not completely clear.

A final and specific limitation is highlighted in Ford et al. (2017), a body of work that looked at the application of the MLP and ECF in the adoption of PV (quite like the theoretical scope of the book): the ECF does not adequately address how landscape and niches shape the regime within which the behaviour related to PV adoption occurs.

But despite its assumptions and limitations, the ECF was used to define the mainstream culture in the socio-technical regime of the MLP as a dominant energy culture fundamentally made up of norms, practices and material culture together with their wider external influences; material culture equates to the 'have', norms to the 'think', and practices to the 'do' (Stephenson, 2012).

External influences, material culture, norms and practices

The first element that will be looked at based on Figure 3.2 is its external influences. Energy cultures are influenced by a wide range of external factors (Stephenson et al., 2015). These influences reside in the space beyond the pseudo-boundary (dotted perimeter) in Figure 3.2, and Stephenson et al. (2015) outline that this pseudo-boundary is particularly useful in making the theoretical distinction between the internal and external aspects of the ECF.

External influences make up the contextual environment that energy cultures exist in (Stephenson et al., 2015). They are beyond the control of the energy culture's actors though external influences affect these actors' practices, norms and material culture (Stephenson et al., 2010). Stephenson et al.'s (2015) work also identifies with the fact that external influences can lock-in patterns of behaviour – and Ford et al. (2017) link this to when the landscape and regime-forces of the MLP support the incumbent structures. Examples of external influences are standards, subsidies, prices and social marketing campaigns (Barton et al., 2013).

With this appreciation for the external environment, material culture, norms and practices can now be discussed. Strengers (2010) describes material infrastructure as physical items such as objects, technologies, infrastructures or even systems of provision which make up an integral part of everyday life. Energy technologies as material culture represent physical and tangible assets that play a role in energy use (Stephenson et al., 2015); they unite the realms of doing and knowing (Drucker, 1961). Examples of material culture are heating technologies like electric heaters, furniture, housing structures, single-glazed windows and low income (Wallenborn, 2008; Sarrica et al., 2016; Stephenson et al., 2016).

Norms are shared beliefs about how people should behave contextually (Stephenson et al., 2015). They are important because the beliefs about energy affect how it is used and vice versa (Strauss, Rupp & Love, 2013). Hopkins and Stephenson (2014) and Sarrica et al. (2016) further state that norms are expectations and aspirations, and both are influential in behavioural change. Examples of norms are the expectations of comfort or even maintaining traditions learnt growing up (Sarrica et al., 2016; Stephenson et al., 2016).

Practices are analytical units that give insight into how daily life persists and are a mixture of perceptions and memories where the latter are products of and produce practices (Wallenborn & Wilhite, 2014). Practices include both routinized and non-routinized actions (Sarrica et al., 2016). Gram-Hassen (2008) distinguishes between both by stating that routines are taken for granted, lack reflection and based on practical experience. Examples of practices are temperature settings, technology maintenance, number of heated rooms, hours of heating or putting on more clothing when feeling cold (Sarrica et al., 2016; Stephenson et al., 2016).

Practices also do not exist as isolated elements but tend to cluster over time and space (Torriti, 2017). Stephenson et al. (2010, 2015) believe that clusters of practices will emerge from a population to create energy cultures which can be used to better design policy interventions – hence the value of understanding local cultures (Wüstenhagen et al., 2007).

Energy subcultures and niche-level culture

Stephenson et al.'s use of the term 'energy cultures' targets combinations of norms, material culture and practices within an external context that form clusters which can separate one energy culture from another. These clusters arise because such cultures are not homogenous or equally distributed within a social system (Spencer-Oatey, 2012; Barton et al., 2013).

Within Stephenson et al.'s work they identify nested energy cultures where the external influences of one energy culture can be part of another energy culture at a higher scale. Figure 3.3 illustrates this conceptualization

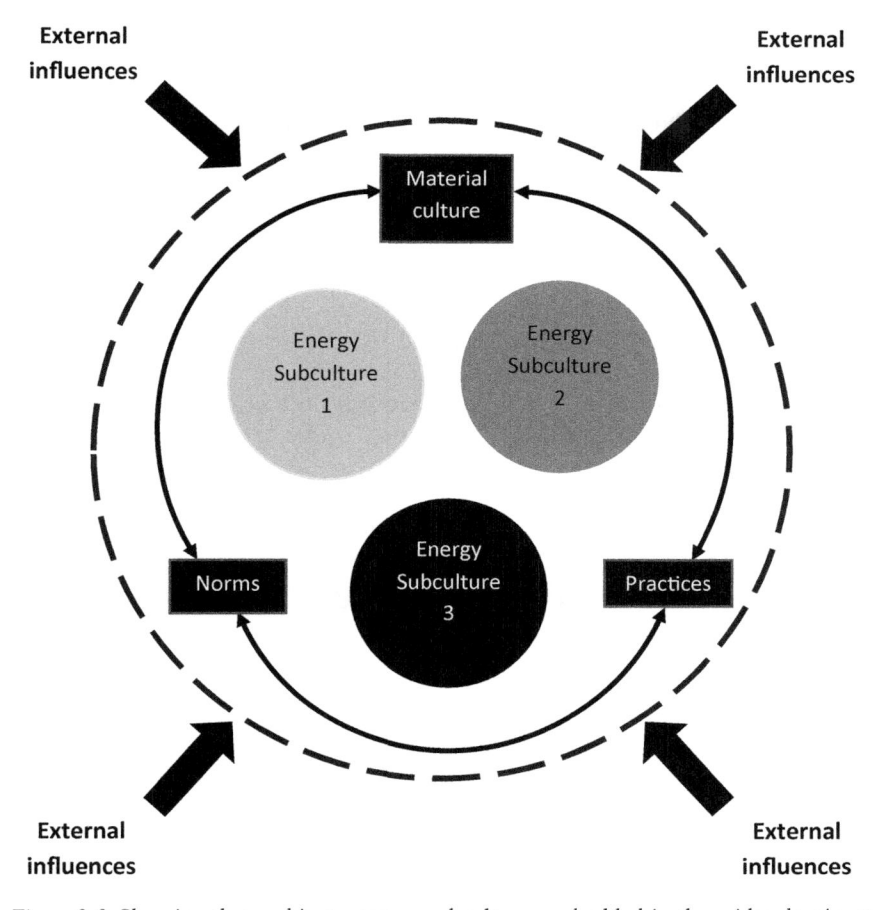

Figure 3.3 Showing three arbitrary energy subcultures embedded in the wider dominant energy culture.

and shows that their model essentially extends consideration to the existence of energy subcultures because of these nested energy culture dynamics.

The formation of energy subcultures suggests that the emergence of new habits usually coincides with a shift in energy culture (Stephenson, 2012). Recalling the importance of adoption to mainstreaming (see Chapter 2), the adoption of technologies is also a key process that can change material culture (as well as norms and practices) according to Stephenson et al. (2010) – and a change in one element of the ECF may create a cascade in the others (Barton et al., 2013; Hopkins & Stephenson, 2014). Therefore, adopting new material culture (like the household SWHs and PV in this book) can lead to an energy culture shift.

This is important to consider because the ECF is used to define the mainstream culture of the regime, and Ford et al. (2017), in one of the bodies of research that looks at the application of the MLP and ECF together, state that changes in energy cultures can reach a critical mass (perhaps linked to the 16% pseudo-mainstreaming threshold from Chapter 2) and can change the nature of the regime. So, this is strongly suggestive of an energy transition and the potential role(s) of 'culture' in it because of the change in regime-orientation.

Based on this it is reasonable to suggest that enough adoption of a niche-innovation can catalyse an energy culture shift (and an energy transition). Therefore, the energy subculture notion is quite like niche-level culture in the MLP. So, by combining the MLP and ECF, 'culture' can be approached as a holistic landscape, regime and niche-level construct.

Figure 3.4 illustrates these dynamics and links the energy subcultures and mainstream energy culture to the landscape, regime and niche planes of the MLP. It shows how niche energy subcultures are integrated into the regime's mainstream energy culture, as well as how the landscape affects these by giving a more three-dimensional view of Figure 3.3.

Linking the multi-level perspective and energy cultures framework

This is the second of two theoretical chapters. So, to complete the discussions related to the definition and roles of 'culture' in energy transitions started in the previous chapter, it is worth explicitly outlining how the two theories/frameworks used in the book are linked; this brief section will present nine theoretical linkages between the MLP and ECF.

The first is that based on 'culture' having both proactive and reactive roles in energy transitions; this creates a window for looking at what it means as a socio-technical feature of the MLP.

Second, 'culture' is explicitly included in the MLP as part of the regime. This acted as the next step in understanding 'culture' in energy transitions by using the ECF's norms, material culture, practices and their external influences to explore it as the energy culture of 'the mainstream'.

The third link is that 'culture' transcends the socio-technical landscape and socio-technical regime of the MLP. Given its complexity and slow rates of change, 'culture' is a landscape-level phenomenon. However, given its explicit placement in the regime as an endogenous mainstream element, 'culture' is also a feature of regimes.

The fourth link relates to the existence of energy subcultures and the adoption of niche-innovations. The work on the ECF that outlined the existence of clusters of features which form an energy culture also showed that they can be nested within a wider energy culture. This therefore indicates that there can be energy subcultures related to niches in the MLP.

Energy transitions and mainstream energy cultures 59

Landscape and regime-level factors are external influences on the dominant energy culture as well as energy subcultures' practices, norms and material culture

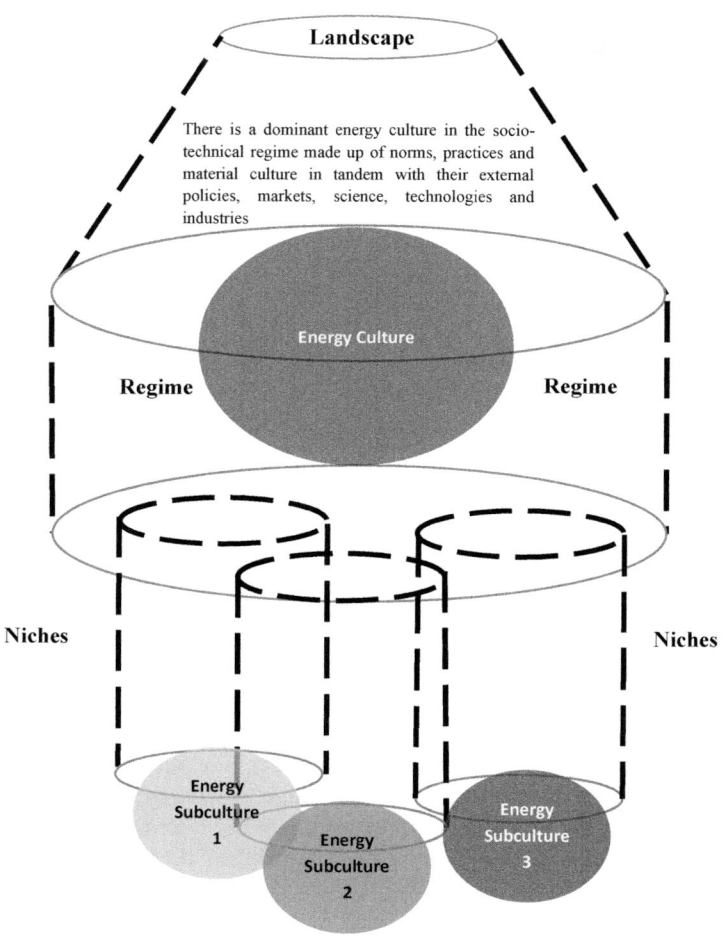

Niches have energy subcultures that collectively integrate into the dominant energy culture where clusters of norms, practices and material culture define the respective subcultures, and external influences on one subculture can be in another subculture

Figure 3.4 Showing 'culture' transcending the three planes of the MLP.

So with respect to the adoption of niche-innovations, the acquisition of new technologies as material culture can be expected to have accompanying changes in norms and practices (based on the ECF's theoretical framing) to create an energy subculture (innovation-adopters) within the larger mainstream energy culture of the regime (adopters and non-adopters).

Fifth, the MLP's landscape and the ECF's external influences are comparable. Given that landscape-level forces are removed from the regime and its actors, it is like the nature of external influences which are also removed from the ECF's core of material culture, norms, practices, and the members of the energy culture. However, regime-forces can also be external influences on an energy culture (Ford et al., 2017). This therefore once again demonstrates the lack of a clear boundary between landscape and regime-level 'culture'.

The sixth link is the connection that 'culture' has with 'technology' as part of the regime. Technology in the MLP is like the ECF's material culture (though technology in the MLP can also refer to the wider construct of technological contexts). Additionally, a common representation of changes to material culture in the ECF is the adoption of new technologies and this fits with the adoption of technological innovations that is pivotal in the MLP.

The seventh linkage is the focus on using the ECF to better design policy interventions, and the occurrence of 'policy' as an explicit feature of the regime in the MLP. This means that there is an opportunity to use the ECF to explore policies in the regime which help with looking at the potential interactions between 'policy' and 'culture' in the book.

The eighth link relates to time. The MLP shows consideration for time. However, the ECF as a model does not explicitly consider time as feature. Therefore, the ECF can be thought of as a tool for exploring snapshots of energy cultures during an energy transition: in theory comparing energy cultures' snapshots at two different points in time during an energy transition can provide insight into the role(s) of 'culture' in the MLP.

The final linkage is that of the interconnectedness between the elements in the MLP's regime (the plane that undergoes an energy transition), as well as those of the ECF. The regime's six elements are part of its co-evolutionary nature so a change in one element would affect one or more of the others (see Chapter 2). Likewise, in the ECF, a change in either the norms, material culture or practices of an energy culture is likely to affect the other elements. Given that the ECF was used to define 'culture' in the regime, this means that a change in any of the regime's other five elements would affect the mainstream energy culture's norms, practices and material culture – and Ford et al. (2017) touch on this link.

Summary

In its broadest sense, 'culture' is a highly interconnected complex of beliefs, values, attitudes, behaviours, norms and experiences with material objects that are learnt through socialization and that are shared by at least two persons. 'Mainstream culture' is the culture that is dominant within a society and is widespread enough to be held by the majority.

'Mainstream culture' in the context of energy transitions and the MLP refers to the culture that is integrated into the socio-technical regime and which influences and is influenced by the regime's policies, technologies, markets, science and industries. This means that in the case of the small, tropical islands being studied in this book, their mainstream cultures are held by the respective populations residing on the different islands.

Though theoretically the mainstream culture exists in the regime, relative to the MLP, 'culture' is not only limited to the regime but is complex enough to co-exist in the socio-technical landscape as well as specific enough to exist in niches. Therefore, the broader culture transcends the three planes of the MLP.

The role(s) and definition of 'culture' in the MLP were not well-articulated or understood in the energy transitions literature reviewed. So, in order to approach the notion of 'culture' with respect to energy, the ECF was used. It approaches 'culture' as an assemblage of norms, practices, material culture and their external influences referred to as an energy culture. Therefore, the socio-technical regime will harbour the mainstream energy culture in the MLP. Mainstreaming refers to an innovation evolving out of its niche(s) to become part of the regime so this suggests that there can be energy subcultures associated with niches and which are subsumed within the more dominant mainstream energy culture.

References

Arnesen, M. (2013). *Saving Energy Through Culture: A Multidisciplinary Model for Analyzing Energy Culture Applied to Norwegian Empirical Evidence* (Master's thesis, Norwegian University of Science and Technology, Trondheim, Norway). Retrieved from https://core.ac.uk/display/52098622.

Barton, B., Blackwell, S., Carrington, G., Ford, R., Lawson, R., Stephenson, J., Thorsnes, P. & Williams, J. (2013). *Energy Cultures: Implications for Policymakers.* Research Report, Centre for Sustainability, University of Otago, Dunedin New Zealand. Retrieved from www.otago.ac.nz/csafe/research/energy/otago055630.pdf.

Berkhout, F., A. Smith & A. Stirling. (2004). Socio-technical Regimes and Transition Contexts. In B. Elzen, F. Geels & K. Green (Eds.), *System Innovation and the Transition to Sustainability: Theory, Evidence and Policy* (pp. 48–75). Cheltenham: Edward Elgar. Retrieved from https://doi.org/10.4337/9781845423421.00013.

Berkowitz M. K. & Haines, Jr., G. H. (1984). The Relationship Between Relative Attributes, Relative Preferences and Market Share: The Case of Solar Energy in Canada. *Journal of Consumer Research*, *11*(3): 754–762. Retrieved from www.jstor.org/stable/2489065.

Broesch, J. & Hadley, C. (2012). Putting Culture Back into Acculturation: Identifying and Overcoming Gaps in the Definition and Measurement of Acculturation. *The Social Science Journal*, *49*(3): 375–385. Retrieved https://doi.org/10.1016/j.soscij.2012.02.004.

Chatterton, T. (2011). *An Introduction to Thinking about 'Energy Behaviour': A Multi Model Approach*. London: Department of Energy and Climate Change. Retrieved from www.gov.uk/government/uploads/system/uploads/attachment_data/file/48256/3887-intro-thinking-energy-behaviours.pdf.

Daghfous, N., Petrof, J. V. & Pons, F. (1999). Values and Adoption of Innovations: A Cross-cultural Study. *Journal of Consumer Marketing*, *16*(4): 314–331. Retrieved from https://doi.org/10.1108/07363769910277102.

Drucker, P. F. (1961). The Technological Revolution: Notes on the Relationship of Technology, Science and Culture. *Technology and Culture*, *2*(4): 342–351. Retrieved from www.jstor.org/stable/3100889.

Dubois, B. (1972). A Cultural Approach to the Study of Diffusion and Adoption of Innovations. In M. Venkatesan (Ed.), *Proceedings of the Third Annual Conference of the Association for Consumer Research* (pp. 840–850). Chicago, IL: Association for Consumer Research. Retrieved from http://acrwebsite.org/volumes/12053/volumes/sv02/SV-02.

Feng, G. & Mu, X. (2010). Cultural Challenges to Chinese Oil Companies in Africa and Their Strategies. *Energy Policy*, *38*(11): 7250–7256. Retrieved from https://doi.org/10.1016/j.enpol.2010.08.001.

Ford, R., Karlin, B. & Frantz, C. (2016). *Evaluating Energy Culture: Identifying Validating Measures for Behaviour-based Energy Interventions.* Paper presented at the International Energy Policies and Programmes Evaluation Conference, Amsterdam, The Netherlands. Retrieved from https://ora.ox.ac.uk/objects/uuid:88323e7b-0a78-44f8-b0ec-d50078b30f88.

Ford, R., Walton, S., Stephenson, J., Rees, D., Scott, M., King, G., Williams, J. & Wooliscroft, B. (2017). Emerging Energy Transitions: PV Uptake Beyond Subsidies. *Technological Forecasting & Social Change*, *117*: 138–150. Retrieved from https://doi.org/10.1016/j.techfore.2016.12.007.

Foxon, T. J., Hammond G. P. & Pearson, P. J. G. (2010). Developing Transition Pathways for a Low Carbon Electricity System in the UK. *Technological Forecasting and Social Change*, *77*(8): 1203–1213. Retrieved from https://doi.org/10.1016/j.techfore.2010.04.002.

Fryberg, S. & Rhys, R. (2009). *Cultural Models*. Retrieved from www.scribd.com/document/296580946/Fryberg-Cultural-models.

Gallivan, M. & Srite. M. (2005). Information Technology and Culture: Identifying Fragmentary and Holistic Perspectives of Culture. *Information and Organization*, *15*(4): 295–338. Retrieved from https://doi.org/10.1016/j.infoandorg.2005.02.005.

Geels, F. W. & Schot, J. (2007). Typology of Socio-technical Transition Pathways. *Research Policy*, *36*(3): 399–417. Retrieved from https://doi.org/10.1016/j.respol.2007.01.003.

Gram-Hassen, K. (2008). Energy Consumption in Homes: An Historical Approach to Understanding New Routines. In M. Rüdiger (Ed.), *The Culture of Energy* (pp. 180–199). Newcastle: Cambridge Scholars Publishing.

Hargreaves T., Longhurst, N. & Seyfang, G. (2013). Up, Down, Round and Round: Connecting Regimes and Practices in Innovation for Sustainability. *Environment and Planning A*, *45*: 402–420. Retrieved from https://doi.org/10.1068%2Fa45124.

Heatwole, C. A. (2006). *Culture: A Geographical Perspective*. New York: The University of the State of New York, The State Education Department. Retrieved from https://pdfs.semanticscholar.org/6cd1/045a14040566467d06341a1c2d59222bf093.pdf.

Hopkins, D. & Stephenson, J. (2014). Generation Y Mobilities Through the Lens of Energy Cultures: A Preliminary Exploration of Mobility Cultures. *Journal of Transport*

Geography, *38*: 88–91. Retrieved from https://ezproxy-prd.bodleian.ox.ac.uk:4563/10.1016/j.jtrangeo.2014.05.013.

Horta, A., Wilhite, H., Schmidt, L. & Baritaux, F. (2014). Socio-technical and Cultural Approaches to Energy Consumption. *An Introduction: Nature and Culture*, *9*(2): 115–121. Retrieved from https://doi.org/10.3167/nc.2014.090201.

King, G., Stephenson, J. & Ford, R. (2014). *PV in Blueskin: Drivers, Barriers and Enablers of Uptake of Household Photovoltaic Systems in the Blueskin Communities, Otago, New Zealand*. Dunedin, New Zealand: Centre for Sustainability, University of Otago. Retrieved from http://hdl.handle.net/10523/5211.

King, R. (2009). Geography, Islands and Migration in an Era of Global Mobility. *Island Studies Journal*, *4*(1): 53–84. Retrieved from www.islandstudies.ca/sites/vre2.upei.ca.islandstudies.ca/files/ISJ-4-1-2009-RKing.pdf.

Klockner, C. (2013). A Comprehensive Model of the Psychology of Environmental Behaviour: A Meta-Analysis. *Global Environmental Change*, *23*(5): 1028–1038. Retrieved from https://doi.org/10.1016/j.gloenvcha.2013.05.014.

Leonard-Barton, D. (1981). Voluntary Simplicity Lifestyles and Energy Conservation. *The Journal of Consumer Research*, *8*(3): 243–252. Retrieved from http://links.jstor.org/sici?sici=0093-5301%28198112%298%3A3%3C243%3AVSLAEC%3E2.0.CO%3B2-V.

Lockwood, M., Kuzemko, C., Mitchell, C. & Hoggett, R. (2013). *Theorising Governance and Innovation in Sustainable Energy Transitions*. University of Exeter, Exeter: Working Paper No. 1304. Energy Policy Group. Retrieved from http://projects.exeter.ac.uk/igov/wp-content/uploads/2013/07/WP4-IGov-theroy-of-change.pdf.

Loorbach, D. (2010). Transition Management for Sustainable Development: A Prescriptive, Complexity-Based Governance Framework. *Governance: An International Journal of Policy, Administration, and Institutions*, *23*(1): 161–183. Retrieved from https://doi.org/10.1111/j.1468-0491.2009.01471.x.

Lutzenhiser, L. (1992). A Cultural Model of Household Energy Consumption. *Energy*, *17*(1): 47–60. Retrieved from https://doi.org/10.1016/0360-5442(92)90032-U.

Lutzenhiser, L. (2008). *Overview of the Behaviour, Energy and Climate Change Conference Keynote Address 'Setting the Stage: Why Behaviour is Important' for California Senate Legislation Development Related to a California Climate Change Research Institute*. Summary of a presentation given at Portland State University, Oregon, United States of America. Retrieved from http://web.stanford.edu/group/peec/cgi-bin/docs/events/2007/becc/presentations/0T-Setting%20the%20Stage%20-%20Why%20Behaviour%20is%20Important%20(Presentation%20Summary).pdf.

Mirosa, M., Gnoth, D., Lawson, R. & Stephenson, J. (2011). Rationalising Energy-Related Behaviour in the Home: Insights from a Value-laddering Approach. *European Council for an Energy Efficient Economy Summer Study* (pp. 2109–2119). Belambra Presqu'île de Giens, France. Retrieved from www.otago.ac.nz/centre-sustainability/otago055638.pdf.

Neuliep, J. W. (2015). Chapter 2 The Cultural Context. In *Intercultural Communication: A Contextual Approach* (5th ed.) (pp. 45–91). Los Angeles: Sage Publications Incorporated. Retrieved from www.sagepub.com/sites/default/files/upm-binaries/42958_2_The_Cultural_Context.pdf.

Raven, R., Schot, J. & Berkhout, F. (2012). Space and Scale in Socio-technical Energy Transitions. *Environmental Innovations and Societal Transitions*, *4*: 63–78. Retrieved from https://doi.org/10.1016/j.eist.2012.08.001.

Rogers, E. M., Medina, U. E., Rivera, M. A. & Wiley, C. J. (2005). Complex Adaptive Systems and the Diffusion of Innovations. *The Innovation Journal: The Public Sector Innovation Journal*, *10*(3), article 30: 1–26. Retrieved from http://citeseerx.ist.psu.edu/viewdoc/download?doi=10.1.1.442.4184&rep=rep1&type=pdf.

Sarrica, M., Brondi, S., Cottone P. & Mazzara, B. M. (2016). One, No One, One Hundred Thousand Energy Transitions in Europe: The Quest for a Cultural Approach. *Energy Research and Social Science*, *13*: 1–14. Retrieved from https://doi.org/10.1016/j.erss.2015.12.019.

Sasin Tipchai. (2016). [*Untitled illustration of actors in costume*]. Pixabay. Retrieved from https://pixabay.com/photos/actor-bangkok-asia-arts-ancient-1807557/.

Sekiguchi, T. & Nakamaru, M. (2011). How Inconsistency between Attitude and Behaviour Persists Through Cultural Transmission. *Journal of Theoretical Biology*, *271*(1): 124–135. Retrieved from https://doi.org/10.1016/j.jtbi.2010.11.044.

Sheller, M. (2014). Global Energy Cultures of Speed and Lightness: Materials, Mobilities and Transnational Power. *Theory, Culture and Society*, *31*(5): 127–154. Retrieved from https://doi.org/10.1177%2F0263276414537909.

Shove, E. (2010). Social Theory and Climate Change Questions Often, Sometimes and Not Yet Asked. *Theory, Culture & Society*, *27*(2–3): 277–288. Retrieved from https://doi.org/10.1177%2F0263276410361498.

Shove, E. & Walker, G. (2010). Governing Transitions in the Sustainability of Everyday Life. *Research Policy*, *39*(4): 471–476. Retrieved from https://doi.org/10.1016/j.respol.2010.01.019.

Shove, E. & Walker, G. (2014). What is Energy for? Social Practice and Energy Demand. *Theory, Culture & Society*, *31*(5): 41–58. Retrieved from https://ezproxy-prd.bodleian.ox.ac.uk:4563/10.1177%2F0263276414536746.

Smith, A., Voß, J. & Grin, J. (2010). Innovation Studies and Sustainability Transitions: The Allure of the Multi-level Perspective and its Challenges. *Research Policy*, *39*(4): 435–448. Retrieved from https://doi.org/10.1016/j.respol.2010.01.023.

Sovacool, B. K. & Hess, D. J. (2017). Ordering Theories: Typologies and Conceptual Frameworks for Socio-technical Change. *Social Studies of Sciences*, *47*(5): 703–750. Retrieved from https://doi.org/10.1177%2F0306312717709363.

Spencer-Oatey, H. (2012). *What is Culture? A Compilation of Quotations*. GlobalPAD Core Concepts. Retrieved from www.warwick.ac.uk/globalpadintercultural.

Stephenson, J. (2012). *Energy Cultures: The Concepts and its Applications (So Far)*. Presentation given at University College London, London.

Stephenson, J. (2018). Sustainability Cultures and Energy Research: An Actor-centred Interpretation of Cultural Theory. *Energy Research & Social Science*, *44*: 242–249. Retrieved from https://doi.org/10.1016/j.erss.2018.05.034.

Stephenson, J., Barton, B., Carrington, G., Doering, A., Ford, R., Hopkins, D., Lawson, R., McCarthy, A., Rees, D., Scott, M., Thorsnes, P., Walton, S., Williams, J. & Wooliscroft, B. (2015). The Energy Cultures Framework: Exploring the Role of Norms, Practices and Material Culture in Shaping Energy Behaviour in New Zealand. *Energy Research and Social Science*, *7*: 117–123. Retrieved from https://doi.org/10.1016/j.erss.2015.03.005.

Stephenson, J., Barton, B., Carrington, G., Gnoth, D., Lawson, R. & Thorsnes, P. (2010). Energy Cultures: A Framework for Understanding Energy Behaviours. *Energy Policy*, *38*(10): 6120–6129. Retrieved from https://doi.org/10.1016/j.enpol.2010.05.069.

Stephenson, J., Barton, B., Carrington, G., Hopkins, D., Lavelle, M. J., Lawson, R., Rees, D., Scott, M., Thorsnes, P., Walton, S. & Wooliscroft, B. (2016). *Energy Cultures Policy Briefs* (Project Report). Dunedin, New Zealand: Centre for Sustainability, University of Otago. Retrieved from http://hdl.handle.net/10523/7104.

Strauss, S., Rupp, S. & Love, T. (2013). Introduction. Powerlines: Cultures of Energy in the Twenty-first Century. In S. Strauss, S. Rupp & T. Love (Eds.), *Cultures of Energy: Power, Practices, Technologies* (1st ed.) (pp. 10–38). Walnut Creek, California: Left Coast Press. Retrieved from https://doi.org/10.4324/9781315430850.

Strengers, Y. (2010). *Conceptualising Everyday Practices: Composition, Reproduction and Change*. Working Paper No. 6. Carbon Neutral Communities. Centre for Design, RMIT University, Melbourne, Australia. Retrieved from www.academia.edu/2076806/Conceptualising_everyday_practices_composition_reproduction_and_change.

Tharp, B. M. (2012). *Defining 'Culture' and 'Organizational Culture': From Anthropology to the Office*. Haworth. Retrieved from www.thercfgroup.com/files/resources/Defining-Culture-and-Organizationa-Culture_5.pdf.

Torriti, J. (2017). Understanding the Timing of Energy Demand Through Time Use Data: Time of the Day Dependence of Social Practices. *Energy Research & Social Science*, *25*: 37–47. Retrieved from https://doi.org/10.1016/j.erss.2016.12.004.

Wallenborn, G. & Wilhite, H. (2014). Rethinking Embodied Knowledge and Household Consumption. *Energy Research and Social Science*, *1*: 56–64. Retrieved from https://doi.org/10.1016/j.erss.2014.03.009.

Wallenborn, G. (2008). The New Culture of Energy: How to Empower Energy Users? In M. Rüdiger (Ed.), *The Culture of Energy* (pp. 236–254). Newcastle: Cambridge Scholars Publishing.

Walton, S., Doeing, A., Gabriel, C. & Ford, R. (2014). *Energy Transitions: Lighting in Vanuatu* (Project Report). Retrieved from http://hdl.handle.net/10523/4859.

White, L. A. (1943). Energy and the Evolution of Culture. *American Anthropologist New Series*, *43*(3): 335–356. Retrieved from www.jstor.org/stable/663173.

Winther, T. (2013). Space, Time, and Sociomaterial Relationships: Moral Aspects of the Arrival of Electricity in Rural Zanzibar. In S. Strauss, S. Rupp & T. Love (Eds.), *Cultures of Energy: Power, Practices, Technologies* (1st ed.) (pp. 164–176). Walnut Creek, CA: Left Coast Press. Retrieved from https://doi.org/10.4324/9781315430850.

Wüstenhagen, R., Wolsink, M. & Burer, M. J. (2007). Social Acceptance of Renewable Energy Innovation: An Introduction to the Concept. *Energy Policy*, *35*(5): 2683–2691. Retrieved from https://doi.org/10.1016/j.enpol.2006.12.001.

Zion, S. & Kozleski, E. (2005). *Understanding Culture*. Arizona: National Institute for Urban School Improvement, University of Arizona. Retrieved from www.researchgate.net/publication/296486383_Understanding_Culture.

Part II
Beginning the household solar energy transition

4 Agriculture, fossil fuels and electricity

Introduction

This book's journey along the energy transition(s) through residential solar energy technologies in small, tropical islands begins with a look at the first case study, Trinidad – an island that has not mainstreamed solar energy just yet. Trinidad is an atypical energy case study by small island standards because unlike most islands, it is rich in fossil fuel resources. So, Trinidad has a thriving indigenous fossil fuel economy and is an example of an island at the very start of the (solar) energy transition. But this industry and economy did not develop overnight.

Chapter 2 illustrated that global historical energy transitions involved biomass and similar resources transitioning through oil then gas and finally, electricity. Small, tropical islands are not dissimilar especially given the colonial history of many – including the book's case studies. In that regard, the energy resources used during Trinidad's post-1700 sugarcane industry starts the book's empirical narrative.

Sugarcane originated in New Guinea (Crosby, 2006) and was planted by the Dutch from 1520 in the Caribbean (Eccles, 2015, 40). Since that time there have been changes in Spanish, French and British colonial rules (Artana et al., 2007) in Trinidad and wider Caribbean. One reason for this fluctuant sovereignty is that islands were much easier to colonize than continental locations (Bass & Dalal-Clayton, 1995).

Besson (2011) outlines that Tobago was producing sugar before Trinidad and that by the time the industry had been established on the islands, Tobago had almost exhausted its soil fertility. This catalysed an economic downturn which prompted the British to geopolitically join Tobago and Trinidad in 1898 (Artana et al., 2007).

The sugarcane industry had several energy resources associated with it. The first worth mentioning is bagasse. Bagasse is the fibrous remnants left from squeezing the sucrose out of sugarcane and it is an effective boiler fuel (Shupe, 1982). Ethanol may have been another fuel since molasses, another sugar industry by-product, could be used as feedstock for ethanol production (Holland, 2015). However, it is rather uneconomical to use molasses in this way and so it is more readily used as livestock feed (Shupe, 1982).

Elias and Victor (2005), Crosby (2006) and Chivers (2015) outline that the metabolic power of draft animals was an energy resource in agrarian economies. For example, Crosby (2006) states that animals like horses markedly improved the speed and power available for human activities. But highlighting metabolic energy in the Caribbean sugar economy also includes its humans because sugarcane cultivation was labour-intensive.

The Caribbean is one of the most apt examples of the various phases of colonialism in the world (Japtok, 2000). In Trinidad's case, its sugarcane economy had many human resource inputs through colonial policies such as the African slave trade from the late 1700s until 1838; East Indian indentureship from 1845 until 1915; as well as European and Chinese immigrants throughout its history (Paria Publishing Company Limited, 2000; Artana et al., 2007; Eccles, 2015, 41; NALIS, 2018).

Another energy resource worth highlighting is charcoal. It became prominent in the 1800s (Bissessarsingh, 2014) and is created by slowly burning wood in almost airless conditions (Crosby, 2006). Eckelmann and Joseph (2003) describe it as a widespread cooking fuel and Wartluft (1984) documents its usage. Charcoal was a popular cooking fuel until the 1960s when LPG became the most widespread cooking fuel in the Caribbean (Wartluft, 1984).

The sugarcane economy's growth was sustained well into the 1900s and two major developments marked a new era of sugar production: the consolidation of a sugar workers' trade union and the establishment of Caroni 1975 Limited as the nation's sugar-producing conglomerate (Eccles, 2015). However, the discovery, extraction and use of oil slowly shifted Trinidad's economy and culminated with the end of the sugar industry in 2003 (Thomas, 2009; Eccles, 2015); the sugar economy of the 19th century began transitioning into the oil economy of the 20th (Artana et al., 2007).

Energy in the oil economy of the late 1800s and early 1900s

Trinidad has one of the oldest oil industries in the world (Sergeant, Racha & John, 2003; Marzolf, Cañeque & Loy, 2015). Its petroleum developments had their beginnings in 1595 when Sir Walter Raleigh caulked his ships with asphalt (Mining Maven, 2013) from the world's largest asphalt deposit, the Pitch Lake. The use of pitch as a resource positioned it as a rudimentary precursor to oil as Raleigh took samples back to Britain – which can be thought of as a metaphor and foundation for the petroleum export industry that was to come in the future (Mining Maven, 2013).

The data from an 1855 geological survey led to interest in extracting the oil from pitch and there were attempts to distil its oil in 1857 (Mining Maven, 2013; Persad & Archie, 2016). Kerosene, however, was distilled from the pitch between 1860 and 1865 (Woodside, 1981, 1). The PAT (1952), MEEBI (1979), Besson (2011), Shaw (2011) and Bissessarsingh (2014) suggest that this pitch together with paraffin and lignite would have been amongst the first hydrocarbon-based energy resources in Trinidad.

The first oil well in Trinidad was drilled in 1857 but oil was discovered in 1866 (Sergeant, Racha & John, 2003; McGuire, Poveda & Raphals, 2007; Besson, 2011; Mining Maven, 2013; Espinasa & Humpert, 2016). The MEEBI (1979) asserts that this period marked the 1st phase of the nation's petroleum history. The MEEBI (1979) continues by stating that 1901 to 1907 was the 2nd phase where the southern parts of Trinidad became the exploratory target of the first oil companies.

During the 3rd phase, commercial oil production began in 1907/1908, was first refined in 1912 and was shipped in 1913 (McGuire, Poveda & Raphals, 2007; Thomas, 2009; Braveboy-Wagner, 2010; Jobity, 2013; Espinasa & Humpert, 2016). This period solidified the Barrackpore, Tabaquite and Point Fortin areas of southern Trinidad as the earliest production sites (MENR, 1984). This phase ended in 1914 and brought commercial success through these Southern explorations, leading to oil exportation to Great Britain. After this 3rd phase, the nation's economic vulnerability to external shocks was illustrated through the World Wars (WW).

Between 1914 and 1919, WWI's global impact marked the 4th phase of the nation's petroleum history (MEEBI, 1979). During WWI, Trinidad's petroleum economy was a key asset of the British military forces since it was a valuable refuelling station for the Royal Navy – and Thomas (2009) outlines that local oil production was ramped up when the Navy converted from coal to oil.

In 1916 pipelines were constructed to connect the three aforementioned Southern production sites (analogous to the industry's consolidation) (MEEBI, 1979). But this was not enough to mitigate the War's economic repercussions because it limited the industry's exploration and shifted its focus to oil-refining and maximizing existing wells' production (MEEBI, 1979). The MEEBI's account of the energy history continues with the 1919 to 1924 post-war growth and the industry's stability thereafter until 1930 due to drilling innovations and industrial learning.

In the 1930s and 1940s the nation experienced economic repercussions when the global price of oil dropped due to significant finds in Texas and Oklahoma (MENR, 1984). Gately (1986), Mining Maven (2013), Baumeister and Kilian (2016) and Baffes et al. (2015) outline that the price plummet continued until 1973 and resulted in an industrial contraction globally. But more optimistically, this period marked the exploration of Northern Trinidad.

After 1940, Trinidad's industry hit another major stumbling block with WWII. As with WWI, oil-refining was prioritized due to its military importance and oil exploration was once again limited. However, this time so, too, was oil production (MEEBI, 1979). Though the MENR (1984) states that a marginal amount of the nation's reserves was exploited at this stage, there was a decline in hydrocarbon production during the late 1900s (MENR, 1984; MoE, 1987; Mining Maven, 2013).

Local oil-refining capacity has historically outstripped oil production (McGuire et al., 2007) so the War's impact on oil production together with this refining-production gap opened the gates for oil importation; by 1950 there were over ten million barrels of oil imported (MEEBI, 1979) and importation continued well into the 1960s (Krausmann, Richter & Eisemenger, 2014).

Energy in the natural gas economy of the late 1900s and 2000s

Natural gas has been traditionally viewed by some as a poorer relative of oil (Brinkworth, 1985), even though both are usually found in association with each other (McMullan, Morgan & Murray, 1977). Trinidad's commercial gas production began from 1946 (McGuire, Poveda & Raphals, 2007). The declining production from the terrestrial fields that carried the industry in the early 1900s prompted the existing oil companies to invest in offshore strategies in Western Trinidad after 1950, and significant commercial discoveries were made in 1955 (MEEBI, 1979; McGuire, Poveda & Raphals, 2007; Russell & Bududass, 2012).

Gas which was originally flared was also diverted and used for downstream activity (Espinasa & Humpert, 2016). This gave shape to the gas economy and a classic example was the introduction of gas for power generation in 1953 (Espinasa & Humpert, 2016) which marked the beginning of a natural gas-powered electricity sector.

The downstream energy sector developed on this 'new' energy (re)source and industrialization led to petrochemical industries that used natural gas as a feedstock as well (Thomas, 2009); the advent of the petrochemical sector started the nation's 2nd phase of industrialization after petroleum refining (MENR, 1984).

The offshore gas searches expanded to the east coast of Trinidad in the 1960s but most notably the nation gained independence from the British in 1962 and the Ministry of Petroleum and Mines was consolidated (Espinasa & Humpert, 2016). The 1970 to 1979 post-independence period resulted in a modified energy system due to an oil boom during this time (Artana et al., 2007; Moya, Mohammed & Sookram, 2010) which was the result of an increase in hydrocarbon prices (Dorf, 1984) as well as discovery of significant gas reserves off Trinidad's Northern coast in 1971 (Espinasa & Humpert, 2016).

The high hydrocarbon prices were a result of the OPEC's activity from 1975. The OPEC is a group of prominent oil-exporting countries that coordinates petroleum policies amongst its member countries, in order to negotiate equitable and stable prices for petroleum producers; efficient, economic and uninterrupted supplies to consumers; and an equitable return on capital for investors (OPEC, 2019).

McVeigh (1984) and the OPEC (2016) state that during this period (the 1970s), the OPEC became more internationally prominent as its members (countries like Venezuela and Saudi Arabia for example) consolidated their domestic oil economies, and events such as the Arab oil embargo in 1973 and the Iranian Revolution of 1979 caused prices to rise. But the OPEC is a cartel, so its policies strongly affect oil markets and the 1979 oil price high was maintained by the OPEC reducing its supply for the next six years (Baffes et al., 2015; Baumeister & Kilian, 2016).

Amid this global context, a major local development in the 1970s was the institution of the Petroleum Production Levy and Subsidy Act (PPLS) (Li & Canetti, 2016; Khan & Khan, 2017). This is worth digressing from the historical economic narrative for a bit because it has serious implications for the mainstreaming of solar energy on the island. de Moor (2001), Iwaro and Mwasha (2010) and Kitson, Wooders and Moerenhout (2011) describe a subsidy as an intervention that keeps consumer prices below the market's, producer prices above the market's or reduces the costs incurred by either party.

Borlick and Wood (2014) explain that the purpose of a subsidy is to incentivize the system to adopt a favourable public policy. However, the use of 'desirable' offers no evolutionary insight as to its desirability for whom. Trinidad's energy subsidy was instituted to stimulate economic growth but since policies, societies and economies evolve over time, what was desirable then may not necessarily be desirable now.

This is particularly because such public subsidies are causative agents of unsustainable development; are economically inefficient, expensive and inequitable; perpetuate 'business as usual' attitudes; and increase the climate impact of fossil fuel consumption (de Moor, 2001; Myers, 2000; Iwaro & Mwasha, 2010; Li & Canetti, 2016).

Based on Badcock and Lenzen's (2010) work, Trinidad benefits from financial and externality-oriented subsidies which create a series of transfers through market mechanisms (de Moor, 2001). Iwaro and Mwasha (2010) and Greigg's (2011) work further suggest that Trinidad's subsidies are both consumer- and producer-oriented.

De Moor (2001) adds that production is primarily subsidized in industrialized countries whilst developing countries focus on consumer subsidies, and as the oil price drops, producer subsidies increase whilst consumer subsidies decrease and vice versa. Further, de Moor (2001) also states that by keeping prices below that of the international markets through subsidization, governments incentivize over-consumption, and by keeping production prices at the minimum above international prices, encourage resource depletion.

However, one key driver of the PPLS Act coming into force was its national legislative protection for consumer prices so that residents could have the benefits of low and stable prices from the local hydrocarbon wealth (Solaun et al., 2015) – since this subsidy skews end-user prices compared to those in a competitive market (Iwaro & Mwasha, 2010).

Today, the oil companies producing more than 3,500 barrels of oil per day in Trinidad are responsible for standing the cost of the subsidy by contributing 4% of their gross income (Iwaro & Mwasha, 2010; Greigg, 2011; Li & Canetti,

2016; Solaun et al., 2015; Khan & Khan, 2017). Solaun et al. go on to state that any amount of the subsidy greater than these companies' 4% contributions is paid by the Government. The Government's contributions usually come directly from its Consolidated Fund and in recent time has accounted for as much as 71% of the subsidy (Li and Canetti, 2016).

With this appreciation for the subsidy, the energy transitions narrative can continue. The State's industrialization directive to use natural gas led to an increased gas consumption of nearly 8% per annum between 1975 and 1995 (McGuire, Poveda & Raphals, 2007). The increasing economic prominence of gas led to the consolidation of the State-owned NGC responsible for purchasing, selling and transporting gas (McGuire, Poveda & Raphals, 2007; Espinasa & Humpert, 2013; Samuel, 2013; Espinasa & Humpert, 2016).

In 1996 natural gas output surpassed that of oil (on an energy equivalent basis) for the first time, where 142,600 BOE/day of natural gas were produced versus 141,000 barrels of oil (McGuire, Poveda & Raphals, 2007; Espinasa & Humpert, 2016). McGuire, Poveda and Raphals (2007) and Russell and Bududass (2012) additionally state that the gap between the commodities increased even further when the country entered the LNG industry in 1998/1999.

To further contextualize the natural gas impact during this time, Espinasa and Humpert (2016, 7) outline that between 1995 and 2006 the natural gas production annual growth rate of 16% took it from 126,600 BOE/day to 669,400. Production remained stable near these levels and peaked at 746,700 BOE/day in 2010 (versus peak oil production of 240,000 barrels in 1978) (Espinasa & Humpert, 2016, 7). Marzolf, Cañeque and Loy (2015) also state that in the wider 1990 to 2013 period, the annual gas consumption grew by 359%. An example of Trinidad's gas prominence is the LNG industry where in 2012, 56.6% of the nation's gas was used for LNG production and Trinidad was the world's 6th largest exporter of the commodity (Marzolf, Cañeque & Loy, 2015).

Based on the above insights, the energy sector was being restructured from oil to gas between 1999 and 2002 (Espinasa & Humpert, 2016). However, natural gas has been the central resource since 1975 (Moya, Mohammed & Sookram, 2010) because oil production was slowly declining around that time (Espinasa, 2008; Espinasa & Humpert, 2013). So, from the legacy of oil in the pre-independence and early post-independence years, natural gas became the more important part of the economy as the primary energy supply dominated by oil in the 1960s to 1980s shifted to gas (Espinasa & Humpert, 2016).

Artana et al. (2007) state that Trinidad's economy transitioned from oil to oil and gas (which is quite true resource-wise). However, the gas trends continued and matured into today's present state where many such as Thomas (2009) and Khan and Khan (2017) describe the nation as having a gas-based economy.

Residential electricity in today's natural gas economy

Major events do not determine when an energy transition occurs but for the sake of this chapter's discussions, they will be used as proxies. Trinidad's economies have been shaped over decadal to century timeframes as demonstrated by the rough 390-, 150- and 70-year reigns of the sugar, oil, and natural gas economies respectively – which aligns with the energy transitions theory from Chapter 2. The change from oil to natural gas is thought to have been smoother than the transition from sugarcane to oil; the transition from sugarcane to oil may have taken about 140 years whilst the transition from oil to natural gas may have lasted around 50. These gaps mark potential lag times of disruption the national system may have experienced during the transitions.

However, what is of interest is the electricity sector that has evolved in the natural gas economy which the previous economic transitions have led to. It is just one industry in the wider natural gas economy, but it is the scope of the regime that the book focuses on – and specifically so the residential electricity regime. So, to characterize this regime, the energy sourcing, power generation, electricity consumption as well as its regulation and governance will be looked at.

Energy sourcing

Mitchell (2014) outlines that there is an outsourcing of energy in many countries so there is a difference between places that produce, and those that use energy. Unlike most small, tropical islands, Trinidad has significant local oil and gas reserves (Espinasa, 2008). Roughly 7% of Trinidad's energy is imported as oil (for refining) but an estimated 42% of its energy production is exported as oil and gas (Espinasa & Humpert, 2016, 6).

Trinidad's hydrocarbon exports are sold to markets in East Asia (e.g. China, Japan and Korea), the U.S., Caribbean and South America (e.g. Argentina and Brazil) (Snow, 2012; Jobity, 2013; Statistics Section, 2019). So, in contrast to most other islands, international oil price increases create more revenue for Trinidad and vice versa when prices fall; the exports put the sector's annual revenue in the vicinity of US$3.5 billion (Driver, 2014; Singh, 2015) which amounts to 57 to 60% of the Government's revenue (Espinasa & Humpert, 2013, 57; Marzolf, Cañeque & Loy, 2015, 5).

The island's oil and gas companies enter bidding negotiations with the Government once select areas are opened for exploration and production (McGuire, Poveda & Raphals, 2007; Jobity, 2013). Successful bids therefore determine which firms are involved in the regime. Once the natural gas is extracted, it is first processed by the Phoenix Park Gas Processors Limited and then placed under the NGC's control (McGuire, Poveda & Raphals, 2007; Blechinger & Shah, 2011). From here, the nation's sole electricity transmission and distribution agent, the T&TEC, enters the picture.

Power generation

Since the inception of electricity use in 1886 Trinidad's electricity industry had been privatized, but was converted into a State-led, vertically integrated monopoly with the enactment of the T&TEC Ordinance No. 42 1945 (McGuire, Poveda & Raphals, 2007; Samuel, 2013; Solaun et al., 2015; Espinasa & Humpert, 2016). This legislation empowers the T&TEC as the sole legal electricity distributor and transmitter (Samuel, 2013).

The MoE (1987) and McGuire, Poveda and Raphals (2007) add that in time the nation's gas producers were directed to dedicate a portion of their reserves to the T&TEC and so provided at a preferential low price – as a result of the subsidy outlined earlier in the chapter.

In Trinidad the arrangement is such that the T&TEC purchases gas from the NGC for the IPPs to generate power (RIC, 2006; McGuire, Poveda & Raphals, 2007; Blechinger & Shah, 2011). T&T's power is provided by four producers (Trinity Power, TGU, PowerGen, T&TEC) that operate eight power plants (six in Trinidad and two in Tobago) (Marzolf, Cañeque & Loy, 2015; Espinasa & Humpert, 2016).

The power plants in Trinidad are all IPPs and amount to 2,428.7 MW whilst Tobago's are utility-owned and rated at 85.7 MW (Espinasa & Humpert, 2016; 8); Trinidad's largest power generation unit by capacity is one of PowerGen's and is rated at 852 MW whilst the lowest is rated at 225 MW and belongs to Trinity Power (Espinasa & Humpert, 2016; 11). Trinidad and Tobago are also connected by an undersea power cable (Espinasa & Humpert, 2016). It is also worth mentioning that one of PowerGen's plants, rated at 308 MW, has been recently decommissioned.

In order to provide the IPPs with their power generation feedstock, the T&TEC entered long-term PPAs with them. These PPAs determine the wholesale electricity price at which the T&TEC purchases power and examples include PowerGen's 15-year 819-MW PPA from 1994, and Trinity Power's 30-year 215-MW PPA from 1998 (RIC, 2006); though there are different IPPs, McGuire, Poveda and Raphals (2007) state that the agreements are similar.

Whilst the energy sourcing is quite different for Trinidad when compared to most small, tropical islands, one striking commonality is that they are primarily supported by conventional, centralized power generation and the centralized power generation-electricity consumption pathway is a linear one (Baker, 2014).

Electricity consumption

The T&TEC provides electricity to households and generates revenue from these customers' bill payments – their two main public services, that is, electricity transmission and distribution, and the administrative elements involved, for example billing of accounts (RIC, 2002). The previous sections have already detailed the technical delivery of electricity, so it is the latter that is of interest here and particularly so billing and tariffs.

The T&TEC issues bi-monthly bills to its customers. These bills are based on a residential tariff set at US$0.04 per kWh for the first 400 kWh, $0.05 US$ per kWh for the next 600 kWh, and anything above this 1,000 kWh threshold is rated at $0.06 per kWh (Espinasa & Humpert, 2016). The rate is therefore a rising block tariff design where a lower price is essentially charged for basic needs such as lighting and refrigeration and higher charges thereafter (Citizens Advice, 2016).

The key-informant interviews conducted with policy officials in Trinidad suggested that the retail rates are determined by the power generation costs; cost of the natural gas feedstock; the prices stipulated in the PPAs between the T&TEC and IPPs; foreign exchange rates; replacement costs; grid network costs; and rate of return for the invested parties. These retail rates however do not reflect the actual costs of power generation because as already mentioned in the chapter, the costs of the natural gas feedstock, that is, the fuel costs, are subsidized.

The T&TEC procures the natural gas for the IPPs at a preferential low price set by the Government (McGuire et al., 2007) – and nearly US$20 million of the $300-million-dollar subsidy benefits the electricity sector (Iwaro & Mwasha, 2010). This is significant because the fuel costs are passed through directly to the consumers (RIC, 2006); Figure 4.1 illustrates these dynamics.

Therefore, this means that residential consumers consequently pay subsidized rates – which is not atypical since electricity is amongst the most commonly subsidized household energy resources (Dzioubinski & Chipman, 1999) and Governments tend to subsidize it along more than one point in its value chain (Kitson, Wooders & Moerenhout, 2011). This also means that such a subsidy will be hidden in economic and public infrastructures (de Moor, 2001; Iwaro & Mwasha, 2010) when it should instead be transparent (Borlick & Wood, 2014).

Nevertheless, the extent of the subsidy makes Trinidad's electricity quite cheap and, as a result, energy use is consumptive. The 'typical' Trinidadian household uses between 458 and 1,090 kWh (no air conditioning) or between 1,390 and 1,543 kWh (with air conditioning) every month (REC, 2011, 41; Marzolf, Cañeque & Loy, 2015, 62 and 63). These consumption figures also put the monthly electricity costs of such a 'typical' household between US$21 and $59 (no air conditioning) or $76 and $86 (with air conditioning).

Electricity regulation and governance

The previous sections outlined the technical delivery of electricity from the natural gas feedstock and the oil and gas companies supplying it, to the households which consume the electricity the T&TEC transmits and distributes. However, the institutions mentioned thus far are not the only agencies involved.

78 *Starting the household solar transition*

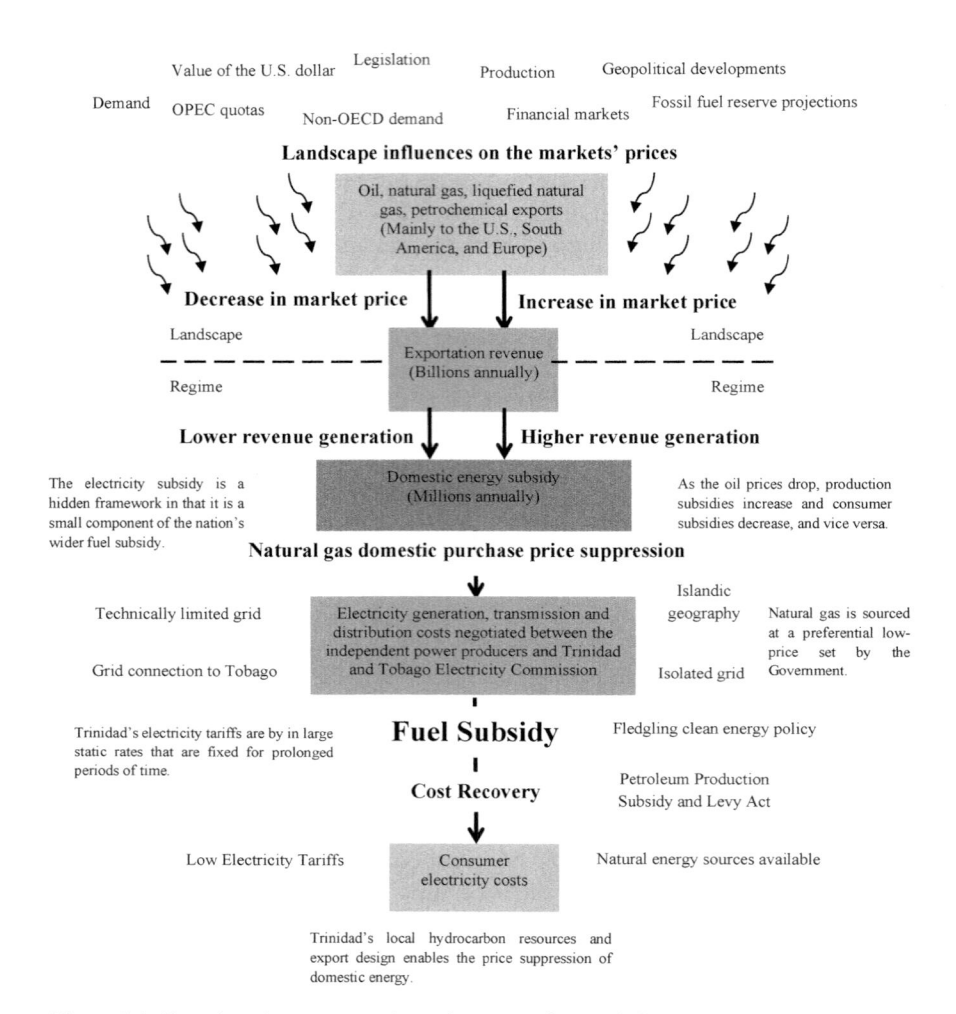

Figure 4.1 Showing the export-oriented nature of Trinidad's energy system.

This section will look at several actors that are part of the wider electricity governance and regulation: the RIC; EID and MPU; MPD; MEEI; EPPD and MEAU; Parliament/Cabinet/Senate; the UWI and UTT; and Sustain T&T as an example of a non-governmental activist group.

In monopolized electricity markets consumers have no choice between suppliers and the supplier often allows the quality of service to depreciate since there is no incentive to maintain it (RIC, 2002). The RIC ensures that good quality and efficient services are being provided by the nation's utilities at equitable costs (McGuire, Poveda & Raphals, 2007). It acts as a watchdog agency over quality (RIC, 2006; Samuel, 2013).

Solaun et al. (2015) state that the RIC regulates the T&TEC and power generators especially as it pertains to rate-setting which involves the T&TEC presenting its case for rate-reviewing; the RIC making its recommendations after public hearings; and final approvals being sanctioned by the Government (McGuire, Poveda & Raphals, 2007). The rates are also set at the best reasonable prices using the methodology provided in the RIC Act (RIC, 2002; MLA, 2015).

The T&TEC and RIC are both regulatory bodies. The T&TEC regulates power generation, electricity transmission and distribution as well as electrical licensing. But when the RIC was consolidated, the regulatory power distribution changed such that quality and rate-setting came under the RIC – and each Commission's regulatory boundaries are determined by their instituting legislations, that is, the T&TEC and RIC Acts respectively. Therefore, the RIC Act determines the nature of the entities which fall under the RIC's jurisdiction.

So relative to these regulatory power dynamics, this is a grey area concerning any future distributed PV because adopting households will be under the jurisdiction of the T&TEC in terms of the generator licensing but any generation tariff for instance may be determined by the RIC – and the latter is only responsible for its scheduled agents and individual generator-households are not featured presently. Therefore, there is the need for legislative revisions.

Not only does the RIC have influence on the T&TEC, but so too do the EID and its line Ministry, the MPU. The MPU is not only responsible for public utilities like electricity, but also others such as water. The Ministry however would have specific policy and legal input into electricity-related protocols, financing and developing projects. The EID is a more specific agency of the Ministry designed to oversee national wiring, electrician advisories and codes of practice for instance. Therefore, given that residential solar energy systems (particularly so distributed rooftop PV) affect the distribution network, the EID and MPU will be part of the technical discussions behind their mainstreaming.

Trinidad's Town and Country Planning Division is a division with vested interests in building development projects and the MPD is its line Ministry. Installing solar systems are construction-related activities so the Division would have advisories for the wider residential sector since it has input in the national developmental agenda which includes environmental issues and to some extent energy issues.

Additionally, one of the most relevant units in the planning and development Ministry is the Economic Development Advisory Board. This Board reports to the Ministerial Council on Economic Development on broad economic diversification strategies aimed at achieving sustainable development which includes seeking out alternative energy initiatives (MPD, 2019). Therefore, it can be expected that the Board will have an interest in mainstreaming residential solar energy.

Trinidad's hydrocarbon history outlined earlier in the chapter implies that the energy Ministry has been and is a powerful institution because it has the political jurisdiction over the country's hydrocarbons. The energy Ministry's mandate is guided by the 1969 Petroleum Act, 1970 Petroleum Regulations and 1974 Petroleum Taxes Act and so the Ministry seeks to promote and manage the nation's energy and mineral resources (McGuire, Poveda & Raphals, 2007; Samuel, 2013; Espinasa & Humpert, 2013; Solaun et al., 2015; MEEA, 2019a). The MEEI is also the NGC's line Ministry (Samuel, 2013) – recalling that the NGC is the nation's sole natural gas merchant. Given the Ministry's oversight on national energy issues, it is intuitive that solar energy development will be under their portfolio.

Energy systems need to transition away from fossil fuel resources to 'greener' options like solar energy and environmental institutions will be part of that transition. So, here enters the EPPD and MEAU. The EPPD is a national Division established to oversee the designing, planning, coordination and implementation of national environmental policies, for example the National Wildlife Policy and National Climate Change Policy. Therefore, its policies will have provisions for energy issues since the energy industry footprint will include implications for wildlife conservation and the emissions associated with global warming and climate change for instance.

Climate change as a standalone policy area is under the purview of the MEAU. The Unit is an entity within the EPPD and specifically looks after the local stakeholder interactions relevant to the climate change policy, carbon reduction strategy and international agreements to which the country is a signatory, for example the United Nations Framework Convention on Climate Change.

The Office of the Prime Minister functions as the highest elected office in the institutional hierarchy on the island but the overarching role of high-level ratification for energy and electricity-related matters is vested with the Parliament/Cabinet/Senate. According to McGuire, Poveda and Raphals (2007) and Samuel (2013) the Parliament/Cabinet/Senate's most important decision-making body with respect to energy is the Cabinet's Standing Committee on Energy that is comprised of ten Government Ministers; technocrats from the energy, finance and planning Ministries; and chairmen and executives of State energy organizations. The Ministry of the Attorney General and Legal Affairs is also involved here because it frames parliamentary bills in law after debate in the Parliament/Cabinet/Senate and would include those policy provisions relevant to energy and electricity.

Whilst the previous institutions were more government-oriented, the UWI, the UTT and Sustain T&T are non-governmental institutions. The UWI and UTT are part of the regime because they are involved in research, development and educational activities linked to energy whether it is on fossil fuels, energy innovations, alternatives like solar energy or environmental sustainability more broadly, for example the UTT's model solar energy house and the UWI's MSc in renewable energy. Sustain T&T is a non-governmental group devoted to sustainability and environmental education through videography and films. The agency is an example of the type of activist groups that would be part of the residential electricity regime.

In addition to the institutions outlined in this section, Solaun et al. (2015) outline that in the wider scheme, the Ministries of Finance, Local Government and Housing and Urban Development would need to be included. The finance Ministry will be particularly important given their role in matters related to fiscal policy as well as taxes and exemptions (McGuire, Poveda & Raphals, 2007) – both of which will affect the economics of residential solar energy technologies. In the case of the Local Government Ministry, their role in facilitating residential solar energy will be through the governance and administration of projects and issuing of support at the municipal level (versus the national scale). Dodman (2008) also suggests that the housing Ministry's Urban Development Corporation of Trinidad and Tobago will play an influential role in planning urban solar energy projects because national issues related to the housing sector are directly under the Ministry's portfolio.

Residential solar energy in the incumbent regime

Resistance from the regime and the barriers to mainstreaming

The theory from Chapter 2 outlined that socio-technical regimes exhibit an adaptative capacity – part of which involves resisting change whether actively through specific and targeted directives and/or developments or passively through legacy-based structures. So, with the detailed background on the evolution of Trinidad's residential electricity regime presented earlier, it is at this junction where the forms of resistance to residential solar energy can be considered since several of the barriers are rooted in the energy history described.

Rogers, Chmutina and Moseley (2012) state that the most significant barriers to renewable energy implementation on small islands are usually financial, institutional and/or political in nature. In Trinidad they include limited electric utility experience with renewables; different ministries coordinating power generation and electricity distribution; lack of renewable energy research and development; lack of fiscal support for energy conservation and alternatives; low public awareness and information access; lack of capital, information and skilled labour; limited technology transfer opportunities; and subsidized electricity (Haraksingh, 2001; Espinasa, 2008; Sharma & Aiyejina, 2010; MEEA, 2012; Solaun et al., 2015; Espinasa & Humpert, 2016; Marzolf, Cañeque & Loy, 2015; Khan & Khan, 2017).

Further, the policy officials interviewed gave in-situ context to the literature's barriers and Table 4.1 shows the various items flagged as barriers in the raw coding structure extracted from the deductive thematic content analysis of the interviews' transcripts. So, based on Table 4.1 and the policy-oriented literature, cost is the key barrier – which is due to the heavily subsidized electricity.

82 Starting the household solar transition

Table 4.1 Showing the deductive coding structure related to the barriers to solar energy implementation in Trinidad

Coding structure			Frequency
Theme/concept	Category	Code	
Barriers to solar energy	Economics	Expensive	22
		Export design of the energy system	6
	Institutions	Lack of policy frameworks and interventions	4
		Lack of supporting legislative instruments	6
		Governmental bureaucracy	13
	Cultural acceptance	Conflicting personal preferences	2
		Lack of public awareness and education	13
		Business as usual culture	7
		Low public engagement	1
	Other	Lack of technical and industrial capacity	5
		Lack of funding opportunities	6

Source: The author.

Note
The frequency is defined as the number of responses that are tagged with the associated code.

The officials also characterized an energy innovation by providing one or more attributes that would give it have a higher chance of being mainstreamed. This was important to see what the description of a 'successful' energy innovation in Trinidad's regime would be (institutionally at least), and how the island's solar energy policies capitalize or can capitalize on these attributes. There are 21 different characterizing attributes and Figure 4.2 shows these; of these, cost-effectiveness is the most prominent.

The above suggests that the economics of household solar energy will be skewed and not as favourable as they could be under competitive market scenarios. For example, recalling that the retail rates are currently US$0.04 per kWh for the first 400 kWh; $0.05 per kWh for the next 600 kWh; and $0.06 per kWh for any amount above this 1,000 kWh threshold, by comparison for instance, distributed residential PV system costs in the Caribbean average US$3.85 per W and ranges from $1.50 in Antigua to $8.00 in Jamaica (GTM Research and Meister Consultants Group, 2015, 10).

However, these rates are not readily seen by householders because it is the upfront capital costs that they face when purchasing solar systems. There is no readily available data (yet) for the standard capital costs of residential solar energy systems in Trinidad because the industry has not been consolidated and the costs of residential installations vary by project. But it would be fair to say that due to the subsidized electricity, as well as low technical capacity

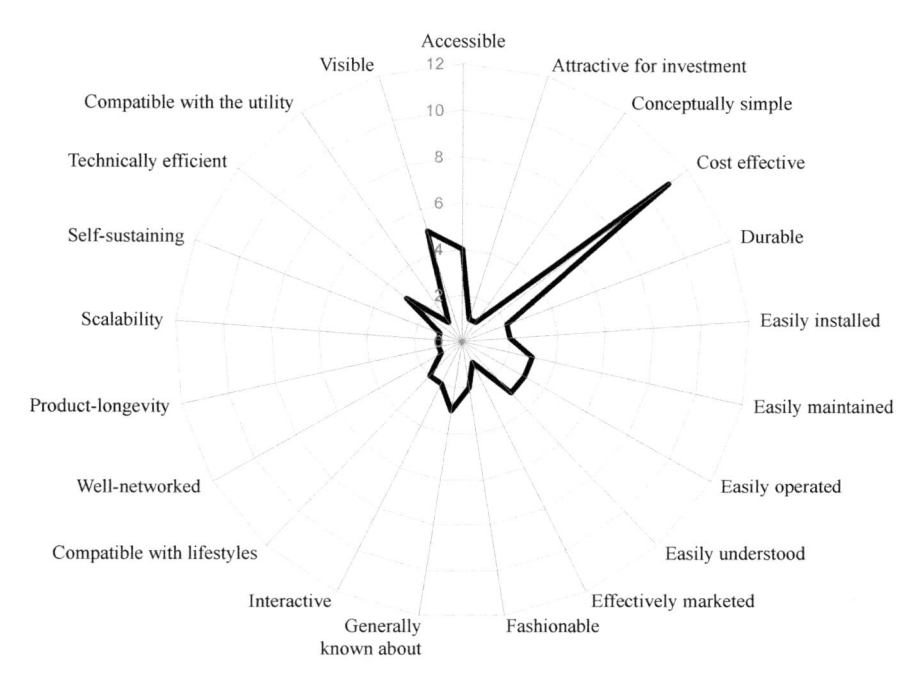

Figure 4.2 Showing the relative significance of the characteristics policy officials believe are important for mainstreaming an energy innovation in Trinidad.

of the local solar energy industry which would affect installation and maintenance costs, the capital costs are expected to be quite high.

Policies as instruments of mainstreaming

Despite the barriers restricting the mainstreaming of solar energy and the consequent stifled market, there are still positive developments which support SWHs and PV's evolution out of their niches: two aspects that will be outlined here are the Government's policy strategies and potential institutional support.

The institutional significance of 'expensive costs' and 'cost-effectiveness' in Table 4.1 and Figure 4.2 respectively is important because policies can change the economics of solar energy and influence investment decisions (Durham, Colby & Longstreth, 1988). They can therefore be influential agents in the household adoption of solar technologies (Bauner & Crago, 2015). However, the future of solar energy can often be dependent on the appropriate government policies.

The policy officials in Trinidad gave the most recent in-situ account of the policy situation. Analysing the data from these interviews resulted in governmental strategies linked to grid-connected solar energy, RETs, legislative reform, tax breaks and stimulating the market. This primary data was coupled with policy-related literature such as the MEEA (2012), NREL (2015), Marzolf, Cañeque and Loy (2015), Khan and Khan (2017), and MEEA (2019b, c, d, e) to create Table 4.2 which shows the current policy environment for solar energy in Trinidad.

Table 4.2 Showing the extent of Trinidad's residential solar energy policy development

Strategy	Technology	Status	Notes
Legislative reform	PV	Pending	The legislations that empower the T&TEC and RIC need to be revised to facilitate the integration of PV.
FiT	PV	Pending	A FiT is being planned to facilitate the integration of distributed PV.
Renewable energy policy	SWHs & PV	Pending	There is a Cabinet-approved Renewable Energy Policy Framework and it has been incorporated into the Draft Policy Green Paper which is being finalized.
RETs	PV	Implemented	There is a prescriptive ambition of 10% of the nation's power generation which must come from renewable energy sources by 2021.
Electric wiring codes	PV	Implemented	The wiring codes applicable to the integration of PV have been revised and published and distributed PV demonstration projects have been set up.
Import duty exemptions	SWHs	Implemented	Components for the manufacture of SWHs are exempt from import duties (MEEA, 2019b).
0-rated VAT	SWHs & PV	Implemented	SWHs and PV panels are free of VAT upon purchase (MEEA, 2019b).
Tax incentives*	SWHs	Implemented	SWH adopters are eligible for a tax credit of 25% the cost of the system up to a maximum cost of USD $1,500 (MEEA, 2019b).
Wear and tear allowances	SWHs & PV	Implemented	An allowance of 150% of the costs incurred on the acquisition of components used in SWH manufacturing, and the acquisition of PV and SWH systems (MEEA, 2019b).

Strategy	Technology	Status	Notes
Technical standards	SWHs	Implemented	There are published standards for the design and installation of SWHs (inclusive of performance, operation and servicing, safety, reliability and durability) (MEEA, 2019e).
Education and public awareness campaigns	SWHs & PV	Implemented	There is a national education campaign to raise the public awareness of the energy sector, renewable energy, energy efficiency and energy conservation (MEEA, 2019d).

Source: The author.

Note
*According to the Imbert (2019, 136), there was a decision to increase the tax incentives for solar water heating to 100% from 1 January, 2020 and this should benefit an estimated 12,000 households.

In this same policy space and in addition to the institutions described earlier on in the chapter, there has been consideration extended towards creating a specialized governmental authority to oversee renewable energy developments. In 2008 the Cabinet agreed on the formation of a Renewable Energy Committee. It consists of nine members from the UWI, UTT, National Energy Corporation, T&TEC and the ministries of energy, trade and industry, public utilities, planning and development, and science and technology (REC, 2011) – which are agencies that have been looked at in this chapter.

The Committee subsequently published the National Framework for a Renewable Energy Policy in 2011 as a guide for the Government's renewable energy plans. Further, the REC (2011) advises the Government to establish the REEEA charged with the duty of developing, assessing, implementing and auditing energy efficiency as well as renewable energy policies and programmes. Based on the Committee's recommendations the REEEA would:

- Develop a register of local renewable energy and energy efficiency businesses, technology providers and experts.
- Create an appropriate regulatory environment for renewable energy and the promotion of energy efficiency and conservation.
- Oversee a national programme of energy audits to determine the present level of energy efficiency, and to identify areas for improvement.
- Institute energy efficiency programmes.
- Oversee the introduction of renewable energy technology such as SWHs and PV to the public.
- Promote public awareness campaigns, and education and training related to renewable energy and energy efficiency.

- Develop incentive programmes for renewable energy technologies.
- Identify and assess proposed renewable energy projects.
- Organize long-term studies and surveys to produce national wind and solar atlases.
- Pursue funding for renewable energy projects from international financial agencies.
- Promote renewable energy businesses.
- Establish working relationships with similar renewable energy authorities and governing bodies in other nations.

This proposed agency is significant to mention because several of the policies that are highlighted in Table 4.2 fit within the recommended remit of the REEEA. The policies are currently coordinated by many governmental institutions (namely the energy Ministry) but an institution like the REEEA could potentially be a more appropriate entity.

Summary

Trinidad's post-1700s sugarcane economy is an example of an agrarian economy, and its main energy resources were likely bagasse, ethanol, the metabolic energy of humans and draft animals, as well as charcoal. However, this economy slowly transitioned during the late 1800s through/with the discovery, extraction and use of oil (in addition to pitch, paraffin, lignite as well as LPG) – and the oil economy's prominence continued into the early 1900s.

But from the late 1900s into 2000s, the discovery, extraction and use of natural gas shifted Trinidad's economy yet again. Whilst oil was and still is a key resource, natural gas became the primary fossil fuel because it is the largest share of the primary energy supply and other downstream industries such as petrochemicals and power generation developed in tandem with its usage.

These shifts are examples of industrial transitions and the island's economy transitioned from using a diversified, low quality mix of energy resources to using fewer, higher-quality energy resources which brought increased economic profitability. These transitions also showed geographic energy sourcing shifts as hydrocarbon production started onshore, then moved offshore and is now in the deep-water territory; all these locations are still productive so contribute to the industry's historical longevity.

There are over 150 years of institutional legacy behind Trinidad's hydrocarbon economy and about 70 of those years are linked to natural gas. These thoughts suggest that the longer a given regime has been in place on an island, the longer it will take for an energy transition to occur because the structures have been locked-in to maintain the regime's incumbency and to resist change, and the more seamless changes are in/to the incumbent regime, the more likely there is a shorter energy transition timeframe involved.

Trinidad's residential electricity sector is a subsector of the electricity industry, which is, in turn, a subsector of the natural gas economy. The residential electricity regime is characterized as being based on large-scale, centralized power generation units which are owned by IPPs and which use natural gas for power generation. The T&TEC is the regime's sole legally authorized electricity transmission and distribution agent and it has long-term PPAs in place with the IPPs. The T&TEC also negotiates with the NGC, the island's designated natural gas institution, to provide the IPPs with the natural gas required for power generation.

The costs of electricity-provisioning, for example fuel costs and fixed costs, are passed on to households and are recovered through the bi-monthly bills that the T&TEC issues. However, given that energy in Trinidad is heavily subsidized, the ultimate costs which are passed on are indirectly subsidized as well so households benefit from cheap electricity.

This is a major barrier to the adoption and subsequent mainstreaming of household SWHs and PV because it means that they are at an economic disadvantage. Other hindrances include the need for legislative reform, low technical capacity, low social awareness and limited research opportunities on energy/electricity as examples. These amongst the other reasons presented in the chapter show why household solar energy has not been mainstreamed thus far in Trinidad.

Nevertheless, there are several policies that have been put in place to support the adoption of solar energy technologies such as tax incentives, electrical wiring codes and RETs. Others include a possible FiT and consolidation of a national renewable energy and energy efficiency authority, the REEEA.

This chapter characterized the starting point as it were of the small-island (solar) energy transition through/with household SWHs and PV by not only highlighting a small, tropical island regime which has not mainstreamed the technologies, but one that is atypical to such islands given its rich fossil fuel heritage. But recalling that the book is interested in the interactions between 'culture' and 'policy', to balance off the more technical and policy-oriented content of this chapter, the next illustrates what the mainstream energy culture of a regime like the one in Trinidad would look like, that is, a pre-SWH mainstream energy culture.

References

Artana, D., Auguste, S., Moya, R., Sookram, S. & Watson, P. (2007). *Trinidad and Tobago: Economic Growth in a Dual Economy*. Inter-American Development Bank. Retrieved from https://sta.uwi.edu/salises/pubs/workingpapers/16.pdf.

Badcock, J. & Lenzen, M. (2010). Subsidies for Electricity-generating Technologies: A Review. *Energy Policy*, 38(9): 5038–5047. Retrieved from https://doi.org/10.1016/j.enpol.2010.04.031.

Baffes, J., Kose, M. A., Ohnsorge, F. & Stocker, M. (2015). *The Great Plunge in Oil Prices: Causes, Consequences, and Policy Responses*. World Bank Group. Retrieved from www.worldbank.org/content/dam/Worldbank/Research/PRN01_Mar2015_Oil_Prices.pdf.

Baker, B. (2014, February 19). How Hawaii is Transitioning from Oil Dependence to Solar Energy. *EcoWatch*. Retrieved from http://ecowatch.com/2014/02/19/hawaii-oil-solar/.

Bass, S. & Dalal-Clayton, B. (1995). *Small Island States and Sustainable Development: Strategic Issues and Experience*. Environmental Planning Issues, 8, Environmental Planning Group, International Institute for Environment and Development, London. Retrieved from http://pubs.iied.org/pdfs/7755IIED.pdf.

Baumeister, C. & Kilian, L. (2016). *Forty Years of Oil Price Fluctuations: Why the Price of Oil May Still Surprise Us*. Center for Financial Studies Woking Paper No. 525, Frankfurt, Germany. Retrieved https://dx.doi.org/10.2139/ssrn.2714319.

Bauner, C. & Crago, C. (2015). Adoption of Residential Solar Power Under Uncertainty: Implications for Renewable Energy Incentives. *Energy Policy*, 86: 27–35. Retrieved from https://doi.org/10.1016/j.enpol.2015.06.009.

Besson, G. A. (2011). Trinidad's Economy. *The Caribbean History Archives*. Paria Publishing Company Limited. Retrieved from http://caribbeanhistoryarchives.blogspot.co.uk/2011/08/trinidads-economy.html.

Bissessarsingh, A. (2014, August 24). The Age of Coal in Trinidad. *Trinidad and Tobago Guardian*. Retrieved from www.guardian.co.tt/lifestyle/2014-08-24/age-coal-trinidad.

Blechinger, P. F. H. & Shah, K. U. (2011). A Multi-criteria Evaluation of Policy Instruments for Climate Change Mitigation in the Power Generation Sector of Trinidad and Tobago. *Energy Policy*, 39(10): 6331–6343. Retrieved from https://doi.org/10.1016/j.enpol.2011.07.034.

Borlick, R. & Wood, L. (2014). *Net Energy Metering: Subsidy Issues and Regulatory Solutions. Issue Brief September 2014*. Institute for Electric Innovation. Retrieved from www.edisonfoundation.net/iei/documents/IEI_NEM_Subsidy_Issues_FINAL.pdf.

Braveboy-Wagner, J. (2010). Opportunities and Limitations of the Exercise of Foreign Policy Power by a Very Small State: The Case of Trinidad and Tobago. *Cambridge Review of International Affairs*, 23(3): 407–427. Retrieved from https://doi.org/10.1080/09557571.2010.484049.

Brinkworth, M. (1985). *Tomorrow's World Energy*. London: The British Broadcasting Corporation.

Chivers, D. (2015). *Renewable Energy: Cleaner, Fairer Ways to Power the Planet*. Oxford, UK: New International Publications Limited.

Citizens Advice. (2016). *Tackling Tariff Design Making Distribution Network Costs Work for Consumers*. Retrieved from www.citizensadvice.org.uk/Global/CitizensAdvice/Energy/Energy%20Consultation%20responses/Tackling%20Tariff%20Design.pdf.

Crosby A. W. (2006). *Children of the Sun: A History of Humanity's Unappeasable Appetite for Energy*. New York, United States of America: W. W. Norton and Company Incorporated.

de Moor, A. (2001). Towards a Grand Deal on Subsidies and Climate Change. *Natural Resources Forum*, 25(2): 167–176. Retrieved from https://doi.org/10.1111/j.1477-8947.2001.tb00758.x.

Dodman, D. (2008). Developers in the Public Interest? The Role of Urban Development Corporations in the Anglophone Caribbean. *The Geographical Journal*, 174(1): 30–44. Retrieved from https://doi.org/10.1111/j.1475-4959.2008.00267.x.

Dorf, R. C. (1984). Managerial and Economic Barriers and Incentives to the Commercialization of Sola Energy Technologies. *Engineering Management International*, 2(1): 17–31. Retrieved from https://doi.org/10.1016/0167-5419(84)90034-6.

Driver, D. (2014, 7 January). What Percentage of Energy Sector Revenue Comes from Oil? *The Energy Chamber of Trinidad and Tobago*. Retrieved from www.ttenergyconference.org/2015/01/what-percentage-of-energy-sector-revenue-comes-from-oil/.

Durham, A., Colby, B. G. & Longstreth, M. (1988). The Impact of State Tax Credits and Energy Prices on Adoption of Solar Energy Systems. *Land Economics*, 64(4): 347–355. Retrieved from www.jstor.org/stable/3146307 (DOI: 10.2307/3146307).

Dzioubinski, O. & Chipman, R. (1999). *Trends in Consumption and Production: Household Energy Consumption*. Discussion Paper No. 6, Department of Economic and Social Affairs, United Nations. Retrieved from www.un.org/esa/sustdev/publications/esa99dp6.pdf.

Eccles, K. E. (2015) The Sugar Heritage Village and Museum Project: Salvaging the History of the Trinidadian People. *Caribbean Library Journal*, 3: 39–49. Retrieved from https://journals.sta.uwi.edu/clj/papers/vol.3/03.KarenEccles.pdf.

Eckelmann, C. & Joseph, I. (2003). *Association of Self-Employed Forest-Workers in Trinidad: A History of Collaboration and Confrontation*. Paper presented at the XII Word Forestry Congress, Québec City, Canada. Retrieved from www.fao.org/docrep/ARTICLE/WFC/XII/0734-C1.HTM.

Elias, R. J. & Victor, D. G. (2005). Energy Transitions in Developing Countries: A Review of Concepts and Literature. Program on Energy and Sustainable Development Working Paper No. 40, Center for Environmental Science and Policy. Stanford, CA: Stanford University. Retrieved from https://pesd.fsi.stanford.edu/sites/default/files/energy_transitions.pdf.

Espinasa, R. (2008). *Prospects for the Oil-Importing Countries of the Caribbean*. Inter-American Development Bank. Retrieved from http://core.ac.uk/download/pdf/6441576.pdf.

Espinasa, R. & Humpert, M. (2013). Trinidad and Tobago. In *Energy Matrix Country Briefings* (pp. 57–60). Inter-American Development Bank. Retrieved from https://publications.iadb.org/en/publication/11241/energy-matrix-country-briefings-antigua-barbuda-bahamas-barbados-dominica-grenada.

Espinasa, R. & Humpert, M. (2016). *Energy Dossier: Trinidad and Tobago*. Inter-American Development Bank. Retrieved from https://publications.iadb.org/en/energy-dossier-trinidad-and-tobago.

Gately, D. (1986). Lessons from the 1986 Oil Price Collapse. *Brookings Papers on Economic Activity*, 17(2): 237–284. Retrieved from https://econpapers.repec.org/article/binbpeajo/v_3a17_3ay_3a1986_3ai_3a1986-2_3ap_3a237-284.htm.

Greigg, K. (2011). *What to Do about the Petroleum Fuel Subsidy in Trinidad and Tobago? A Sustainable Reform of the Fuel Subsidy*. Presentation given at the 2011 Conference on the Economy, University of the West Indies, St. Augustine, Trinidad and Tobago. Retrieved from https://sta.uwi.edu/conferences/11/cote/documents/Day%201/Panel%204%20-%20Energy%20Challenges/What_to_do_with_the_Petroleum_Fuel_Subsidy_Kyren_Greigg.pdf.

GTM Research and Meister Consultant Group. (2015). *Solar PV in the Caribbean Opportunities and Challenges*. Retrieved from www.researchgate.net/publication/324064167.

90 *Starting the household solar transition*

Haraksingh, I. (2001). Renewable Energy Policy Development in the Caribbean. *Renewable Energy*, *24*(3): 647–655. Retrieved from https://doi.org/10.1016/S0960-1481(01)00051-9.

Holland, R. (2015). *Energy in Developing Countries*. Presentation given at the University of Oxford, Oxford, UK.

Imbert, C. (2019). *Budget Statement 2020*. Ministry of Finance, Government of the Republic of Trinidad and Tobago, Port of Spain, Trinidad and Tobago. Retrieved from www.finance.gov.tt/wp-content/uploads/2019/10/BUDGET-STATEMENT-2020.pdf.

Iwaro, J. & Mwasha, A. (2010). Towards Energy Sustainability in the World: The Implications of Energy Subsidy for Developing Countries. *International Journal of Energy and the Environment*, *1*(4): 705–714. Retrieved from www.ijee.ieefoundation.org/vol.1/issue4/IJEE_13_v1n4.pdf.

Japtok, M. (2000). Sugarcane as History in Paule Marshall's 'To Da-Duh, in Memoriam'. *African American Review*, *34*(3): 475–482. Retrieved from www.jstor.org/stable/2901385 (DOI: 10.2307/2901385).

Jobity, R. 2013. *Tertiary Natural Gas Workshop: July/August 2013*. Presentation published by National Gas Company of Trinidad and Tobago https://ngc.co.tt/wp-content/uploads/pdf/NGC_Webinar_The%20Structure_History_and_Role_of_the_Natural_Gas_Industry_2013-08-22.pdf.

Khan, Z. & Khan, A. A. (2017). Current Barriers to Renewable Energy Development in Trinidad and Tobago. *Strategic Planning for Energy and the Environment*, *36*(4): 8–23. Retrieved from https://doi.org/10.1080/10485236.2017.11863769.

Kitson, L., Wooders, P. & Moerenhout, T. (2011). *Subsidies and External Costs in Electric Power Generation: A Comparative Review of Estimates*. International Institute for Sustainable Development. Retrieved from www.iisd.org/library/subsidies-and-external-costs-electric-power-generation-comparative-review-estimates.

Krausmann, F., Richter, R. & Eisemenger, N. (2014). Resource Use in Small Island States: Material Flows in Iceland and Trinidad and Tobago, 1961–2008. *Journal of Industrial Ecology*, *18*(2): 294–309. Retrieved from https://doi.org/10.1111/jiec.12100.

Li, X. & Canetti, E. (2016). *Trinidad and Tobago – Selected Issues*. Washington, DC: International Monetary Fund. Retrieved from www.imf.org/external/pubs/ft/scr/2016/cr16205.pdf.

Marzolf, N. C., Cañeque, F. C. & Loy, D. (2015). *A Unique Approach for Sustainable Energy in Trinidad and Tobago*. Washington, DC: Inter-American Development Bank. Retrieved from www.energy.gov.tt/wp-content/uploads/2016/08/A-Unique-Approach-for-Sustainable-Energy-in-Trinidad-and-Tobago.pdf.

McGuire, G., Poveda, M. & Raphals, P. (2007). *Competition in Energy Markets: Trinidad and Tobago Case Study*. Latin American Energy Organization. Retrieved from http://biblioteca.olade.org/opac-tmpl/Documentos/old0024.pdf.

McMullan, J. T., Morgan, R. & Murray, R. B. (1977). *Energy Resources*. London: Edward Arnold Publishers Limited.

McVeigh, J. C. (1984). *Energy Around the World: An Introduction to Energy Studies: Global Resources, Needs, Utilization*. Oxford, UK: Pergamon Press Limited.

MEEA (Ministry of Energy and Energy Affairs) (2012). *Renewable Energy and Energy Efficiency Policy Trends and Initiatives in T&T*. Government of the Republic of Trinidad and Tobago. Presented at the Third National CDM Capacity Building Workshop,

Port of Spain, Trinidad and Tobago. Government of the Republic of Trinidad and Tobago. Retrieved from http://trinidadandtobago.acp-cd4cdm.org/media/353735/re-ee-policy-trends-initiatives-tt.pdf.

MEEA (Ministry of Energy and Energy Affairs). (2019a). *About Us*. Government of the Republic of Trinidad and Tobago. Retrieved from www.energy.gov.tt/about-us/.

MEEA (Ministry of Energy and Energy Affairs) (2019b). *Renewable Energy and Energy Efficiency Fiscal Incentives*. Our Business. Government of the Republic of Trinidad and Tobago. Retrieved from www.energy.gov.tt/our-business/alternative-energy/renewable-energy-and-energyefficiency-fiscal-incentives/.

MEEA (Ministry of Energy and Energy Affairs) (2019c). *Pilot Projects*. Our Business. Government of the Republic of Trinidad and Tobago. Retrieved from www.energy.gov.tt/our-business/alternative-energy/pilot-projects.

MEEA (Ministry of Energy and Energy Affairs) (2019d). *National Public Awareness Program*. Our Business. Government of the Republic of Trinidad and Tobago. Retrieved from www.energy.gov.tt/our-business/alternative-energy/national-public-awareness-program/.

MEEA (Ministry of Energy and Energy Affairs). (2019e). *Renewable Energy Technical Standards*. Retrieved from www.energy.gov.tt/our-business/alternative-energy/renewable-energy-technical-standards/.

MEEBI (Ministry of Energy and Energy-Based Industries). (1979). *A History of Oil in Trinidad*. Port of Spain, Trinidad and Tobago: Ministry of Energy and Energy-Based Industries.

MENR (Ministry of Energy and Natural Resources). (1984). *Information on The Petroleum Industry of Trinidad and Tobago: Government of Trinidad and Tobago*. Port of Spain, Trinidad and Tobago: Government of the Republic of Trinidad and Tobago.

Mining Maven. (2013). *Trinidad Oil Report*. Retrieved from www.miningmaven.com/pdf/Trinidad-OilReport.pdf.

Mitchell, T. (2014). *Carbon Democracy: Energy Politics and the Corporate Future*. Presentation given at the University of Oxford, Oxford, UK.

MLA (Ministry of Legal Affairs). (2015). *Regulated Industries Commission Act Chapter 54:73, Act 42 of 1945, Amended by 4 of 2001 (Unofficial Version)*. Laws of Trinidad and Tobago, Port of Spain, Trinidad and Tobago. Retrieved from http://rgd.legalaffairs.gov.tt/laws2/alphabetical_list/lawspdfs/54.73.pdf.

MoE (Ministry of Energy). (1987). *National Energy Policy for Trinidad and Tobago*. Port of Spain, Trinidad and Tobago: Government of Trinidad and Tobago.

Moya, R., Mohammed, A. & Sookram, S. (2010). *Productive Development Policies in Trinidad and Tobago: A Critical Review*. Inter-American Development Bank. Retrieved from https://publications.iadb.org/en/publication/productive-development-policies-trinidad-and-tobago-critical-review.

MPD (Ministry of Planning and Development). (2019). *Economic Development Advisory Board (EDAB)*. Government of the Republic of Trinidad and Tobago. Retrieved from www.planning.gov.tt/content/economic-development-advisory-board-edab-0.

Myers, N. (2000). Sustainable Consumption. *Science*, 287(5462): 2419. Retrieved from https://doi.org/10.1126/science.287.5462.2419.

NALIS (National Library and Information System Authority). (2018). *Slavery and Emancipation in Trinidad and Tobago*. Emancipation Day. Retrieved from www.nalis.gov.tt/Resources/Subject-Guide/Emancipation-Day#tabposition_24761.

NREL (National Renewable Energy Laboratory). (2015). *Energy Snapshot: Trinidad and Tobago*. Energy Transitions Initiative-Islands, United States Department of Energy. Retrieved from www.nrel.gov/docs/fy15osti/64117.pdf.

OPEC (Organization of the Petroleum Exporting Countries). (2019). *Brief History*. Retrieved from www.opec.org/opec_web/en/about_us/24.htm.

Paria Publishing Company Limited. (2000). *The History of Sugar Cane and Rum Part 3*. Retrieved from www.trinbagopan.com/Landofbeginings12.html.

PAT (Petroleum Association of Trinidad). (1952). *Trinidad's Oil: An Illustrated Survey of the Oil Industry in Trinidad*. London: Curwen Press.

Persad, K. M. & Archie, C. (2016). *History of Petroleum Exploration in Trinidad and Tobago*. Presentation given at the American Association of Petroleum Geologists (AAPG) and the Society of Exploration Geophysicists (SEG) International Conference & Exhibition, Cancun, Mexico. Retrieved from www.searchanddiscovery.com/pdfz/documents/2016/70231archie/ndx_archie.pdf.html.

REC (Renewable Energy Committee). (2011). *Framework for Development of a Renewable Energy Policy for Trinidad and Tobago: A Report of the Renewable Energy Committee*. Trinidad and Tobago: Ministry of Energy and Energy Affairs. Retrieved from www.energy.gov.tt/wp-content/uploads/2014/01/Framework-for-the-development-of-a-renewable-energy-policy-for-TT-January-2011.pdf.

RIC (Regulated Industries Commission). (2002). *Quality of Service Standards for the Trinidad and Tobago Electricity Commission Draft for Consultation*. Port of Spain, Trinidad and Tobago: Regulated Industries Commission.

RIC (Regulated Industries Commission). (2006). *Electricity Transmission and Distribution Price Control Review 2006–2010 Draft Determination*. Port of Spain, Trinidad and Tobago: Regulated Industries Commission.

Rogers, T., Chmutina, K. & Moseley, L. L. (2012). The Potential of PV Installations in SIDS – An Example in the Island of Barbados. *Management of Environmental Quality, 23*(3): 284–290. Retrieved from https://doi.org/10.1108/14777831211217486.

Russell, A. & Bududass, R. (2012). *T&T's Petroleum Upstream Sector: A View of the Next 50 Years*. Presentation given at the Presented at the 2012 Conference on the Economy, University of the West Indies, St. Augustine, Trinidad and Tobago. Retrieved from https://sta.uwi.edu/conferences/12/cote/documents/AlanRussellandRajeshBududass-TandTPetroleumIndustrySectorTheNext50Years.pdf.

Samuel, H. A. (2013). *A Review of the Status of the Interconnection of Distributed Renewables to the Grid in CARICOM Countries*. Caribbean Renewable Energy Development Programme, Caribbean Community and Deutsche Gesellschaft für Internationale Zusammenarbeit (GIZ) GmbH Germany, Castries, St. Lucia. Retrieved from www.scribd.com/document/252916035/CREDP-GIZ-A-Review-of-the-Status-of-the-Interconnection-of-Distributed-Renewables-to-the-Grid-in-CARICOM-countries-2013.

Sergeant, K., Racha, S. & John, M. (2003). *The Petroleum Sector: The Case of Trinidad and Tobago. Trends, Policies and Impacts, 1985–2000*. Economic Commission for Latin America and the Caribbean, Santiago, Chile. Retrieved from http://citeseerx.ist.psu.edu/viewdoc/download?doi=10.1.1.530.5156&rep=rep1&type=pdf.

Sharma, C. & Aiyejina, A. (2010). *The Requirements and Effects of a RE Portfolio Standard on the Economies of the Caribbean Island Chain: A Case Study of Trinidad and*

Tobago. Paper presented at the International Institute of Electrical Engineers, Green Technologies Conference, Grapevine, Texas, United States of America. Retrieved from https://doi.org/10.1109/GREEN.2010.5453793.

Shaw, T. M. (2011). Energy and CSR in Trinidad and Tobago in the Second Decade of the Twenty-first Century. In J. Sagebien & N. M. Lindsay (Eds.), *Governance Ecosystems: CSR in the Latin American Mining Sector* (1st ed.) (pp. 245–259). London: Palgrave Macmillan. Retrieved from https://doi.org/10.1057/9780230353282.

Shupe, J. W. (1982). Energy Self-Sufficiency for Hawaii. *Science, 216*(4551): 1193–1199. Retrieved from https://doi.org/10.1126/science.216.4551.1193.

Singh, N. (2015, October 3). T&T Earns $21,186 b from Oil and Gas. *Trinidad and Tobago Guardian*. Retrieved from www.guardian.co.tt/news/2015-10-03/tt-earns-21186b-oil-and-gas.

Snow, N. (2012, October 24). Trinidad and Tobago Energy Minister Outlines LNG Export Changes. *Oil and Gas Journal*. Retrieved from www.ogj.com/articles/2012/10/trinidad-and-tobago-energy-minister-outlines-lng-export-changes.html.

Solaun, K., Gomez, I., Larrea, I., Sopelana, A., Ares, Z. & Blyth, A. (2015). *Strategy for Reduction of Carbon Emissions in Trinidad and Tobago, 2040: Action Plan for the Mitigation of GHG Emissions in the Electrical Power Generation, Transport and Industry Sectors*. Port of Spain, Trinidad and Tobago: Government of the Republic of Trinidad and Tobago.

Statistics Section. (2019). *Trinidad and Tobago Country Fact Sheet*. Department of Foreign Affairs and Trade, Australian Government. Retrieved from http://dfat.gov.au/trade/resources/Documents/trin.pdf.

Thomas, S. M. (2009). *Impacts of Economic Growth on CO_2 Emissions: Trinidad Case Study*. Paper presented at the 45th ISOCARP Congress, Porto, Portugal. Retrieved from www.isocarp.net/Data/case_studies/1598.pdf.

Wartluft, J. L. (1984). *Comparing Charcoal and Wood-Burning Cookstoves in the Caribbean*. Virginia: Volunteers in Technical Assistance. Retrieved from http://pdf.usaid.gov/pdf_docs/PNAAU242.pdf.

Woodside, P. R. (1981). *The Petroleum Geology of Trinidad and Tobago*. United States Geological Survey, United States Department of the Interior. Retrieved from https://pubs.usgs.gov/of/1981/0660/report.pdf.

5 Electricity and mainstream energy cultures

Introduction

Chapter 4 showed that Trinidad's economies transitioned from using resources such as biomass, to oil and then natural gas. In that context, energy production and consumption has set the stage for society's relations with its human and non-human elements (Horta et al., 2014) such that the centralized electricity generation that typifies most of the world, including Trinidad as shown in Chapter 4, is mirrored in energy economics, socio-economic values, behaviours and governance (Nolden, 2012).

Electricity is a quintessential part of modern energy systems and is the most pervasive energy carrier in the world. But its introduction, presence and usage are culture-specific (Winther, 2013). Therefore, it is conceivable that residents' interactions with their electricity supply, consumption and payment systems have developed an energy culture since people's habits would have been locked-in through their interactions within the (energy) culture, over history and with their material environments (Wallenborn & Wilhite, 2014).

In that regard, this chapter gives insight into what Trinidad's mainstream energy culture is like based on structured interviews conducted with residents on the island and the inductive thematic content analysis of their transcripts; it is a qualitative sample from a mainstream energy culture in a residential electricity regime which has not mainstreamed solar energy technologies.

What this means then is that the analytical approach is based on the institutional relationships described in the last chapter between households and the utility company, the T&TEC. The T&TEC transmits and distributes the electricity generated by the regime's IPPs, and households consume this electricity and pay the T&TEC for what they consume. These notions suggest that there would be norms, practices, material culture and external influences that are associated with households' interactions with(in) this system – which can be explored as cultural models.

Cultural models are formed when cultural information groups around a specific domain (Broesch & Hadley, 2012); domains help individuals understand and organize their social world (Fryberg & Rhys, 2009). So, by isolating the institutional relationships summarized earlier, households' energy supply, usage and costs can be studied as cultural domains of the regime's mainstream energy culture.

Energy supply

Modern industries and societies are linked through energy production, delivery and consumption (WEF, 2018). But energy supply and demand have often been treated as separate (Sovacool, 2009; Strengers, 2010) when they should instead be connected. All cultures are dependent on sources of energy (Ruotsalainen et al., 2017). So, householders' 'energy supply' was treated as a cultural domain with the expectation that there would be cultural information that would be linked to it as a concept. In that regard, residents were asked about where they believed they got their energy. It is worth mentioning that the phrasing is important because the term 'energy' was used versus 'electricity'. This was done so that any electricity-related answers could come organically, and residents could include non-electrical forms of energy.

The responses directly give insight into the residents' household energy inputs to detail what technical elements from the regime described in Chapter 4 have been assimilated into the incumbent mainstream energy culture. Further, it is useful to consider because Chapter 2 showed that a common (though limited) metric for determining when an energy transition occurs is noting changes in the technical energy mix. So, a thought here is whether such a change could be observable culturally.

Figure 5.1 shows the household energy inputs based on analysis conducted on the residents' responses. As can be seen, the incumbent mainstream energy culture has assimilated electricity and LPG; the 'Other' seen in the Figure is based on one resident viewing light and sound as forms of energy.

Table 5.1 shows the distribution of the raw coding data for the electricity that is seen in Figure 5.1. The key observations are that there is an awareness that hydrocarbons are the fuel sources for power generation and electricity is a domestic energy input. In the case of the first, only two residents' responses connect with norms related to hydrocarbons, for example:

> energy comes from electricity, which comes from [the] T&TEC, which comes from [natural] gas.
>
> (Resident 14)

The other resident stated that the natural gas resources come from offshore and power turbines to generate electricity. These two responses identify with the use of natural gas and this is important to flag because oil could have been easily misconceived as the power generation feedstock even though, technically, Trinidad's power generation is completely driven by natural gas (see Chapter 4).

96 *Starting the household solar transition*

OTHER
Total Sub-code Frequency count = 2
Total Number of Sub-codes = 2

LIQUEFIED PETROLEUM GAS
Total Sub-code Frequency count = 2
Total Number of Sub-codes = 2

ELECTRICITY
Total Sub-code Frequency count = 26
Total Number of Sub-codes = 8

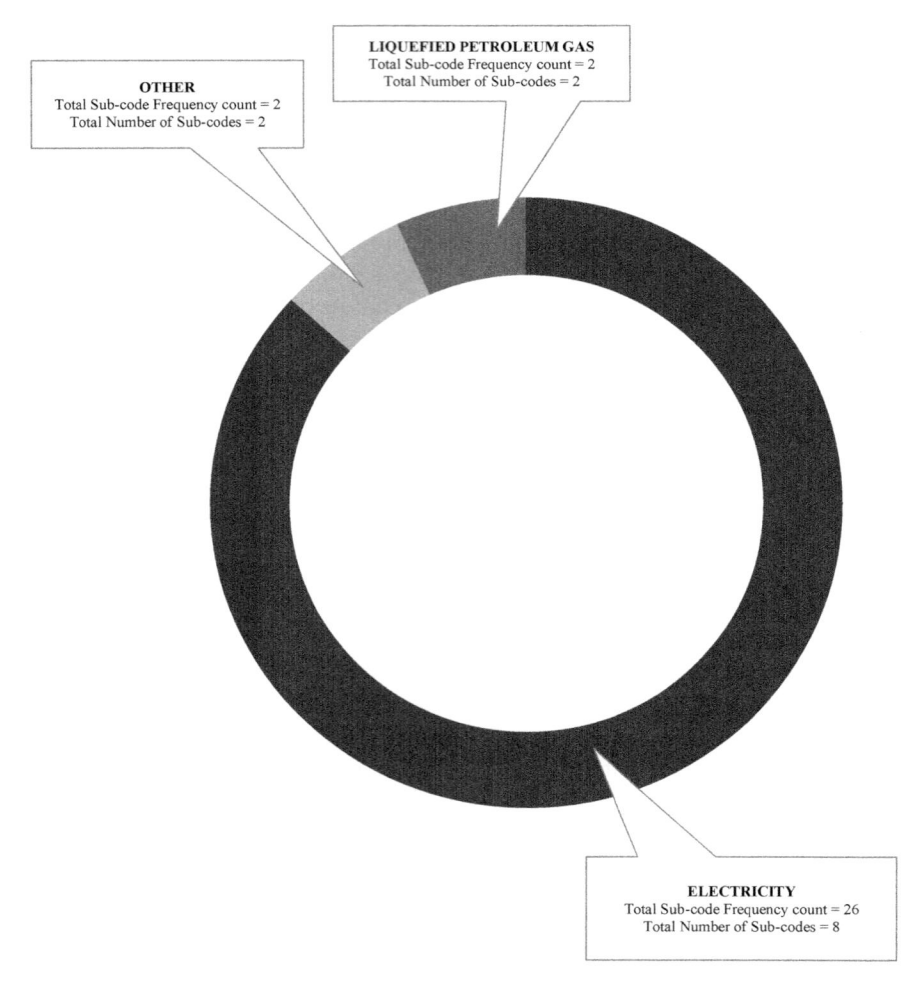

Figure 5.1 Showing the household energy inputs which emerged from the inductive thematic content analysis of the interviews conducted with residents in Trinidad.

The most significant part of the cultural framing of electricity is its simple yet direct reference during the interviews however, for example:

> Well electricity.
>
> (Resident 12)

The references associated with homes using electrical energy is the most frequent in Table 5.1. Linked to this use of electricity is the infrastructure that supplies it, that is, the grid. The grid refers to the infrastructure used to

Table 5.1 Showing the electricity category's coding structure which forms part of the 'energy supply' cultural domain

Coding structure				Sub-code frequency
Category	Sub-category	Code	Sub-code	
Electricity	Electricity is a domestic energy input	A single electric utility company provides electricity to homes	Homes use electrical energy	12
			Transmission and distribution infrastructure	2
			T&TEC provides energy	6
	Hydrocarbons fuel electricity production	Natural gas fuels electricity production	Natural gas is used as the power generation feedstock	2
			Natural gas is sourced from offshore hydrocarbon reserves	1
		Power generation plants generate electricity through turbines	Power generation plant produces electricity	1
			Natural gas-powered turbines generate power	1
			Power generation plant location is known	1

Source: The author.

Note

The frequency is defined as the number of responses that are tagged with the associated sub-code.

generate, transmit and distribute electricity (Lofthouse, Simmons & Yonk, 2015; Citizens Advice, 2016) and is the external network that households are set within where energy generation and transformation connects energy and homes (Wallenborn, 2008); households are a part of a wider material infrastructure system including wires, power stations, appliances and even switches (Strengers, 2010). The data in the table shows that residents were aware (however generically) of the existence of this technical network.

Additionally, the second most frequent reference made by residents is the fact that their energy comes from the T&TEC (see Table 5.1), for example:

Through [the] T&TEC.

(Resident 1)

98 *Starting the household solar transition*

As detailed in Chapter 4, the T&TEC is the electricity transmission and distribution agent in T&T. However, not only is the identification of the T&TEC a unique trait to this energy culture, but it fundamentally means that residents connect with the fact that their supply of electricity comes from a sole electricity utility company.

Further, whilst the analysis and interview recording focused on the explicitly stated information for transcription, interpretation and analysis, it is reasonable to deduce that someone simply stating the company's name, as in the earlier quote, implies that electricity is used in the household. But specifically mentioning the company pinpoints its main function of electricity-provisioning as the answer to the question and not necessarily solely electricity.

The energy used for power generation accounts for nearly 11% of Trinidad's primary energy supply (Espinasa & Humpert, 2016; 6) and residential electricity consumption accounts for nearly 29% of the total electricity consumption (Espinasa & Humpert, 2016; 14). Technical data on the energy (re)source breakdown in an average individual household was not available, however. But in the wider context, electricity is more widely considered to be the most important energy carrier of today (Demirel, 2016).

It is a necessity for living and once accessible becomes a prevalent part of households regardless of income (Hiemstra-van der Horst & Hovorka, 2008; Winther and Bouly de Lesdain, 2013); it has no substitutes (Sovacool, 2009). So, given the pervasive nature of electricity as a commodity, it is expected to be the most widely consumed in Trinidadian households – hence why it is so culturally significant in the sample looked at.

Nevertheless, electricity is not the only culturally salient household energy input. The other resource is LPG. LPG is primarily made up of propane and butane (McGuire et al., 2009; Kojima, 2011; Sepp, 2014; Demirel, 2016). It is produced from oil and natural gas production; oil-refining and gas purification (which yields the propane and butane); upstream transport, refining and storage; and then downstream transport, storage, bottling and retailing (U.S. Department of Energy, 2003; Kojima, 2011; Sepp, 2014; van Leeuwen, de Wit & Smit, 2017).

LPG is the most common cooking fuel in the English-speaking Caribbean (Wartluft, 1984) which would include Trinidad, and the household cylinders commonly referred to as cooking gas, bottled gas, propane or butane are all connotations pointing towards LPG (McGuire et al., 2009; van Leeuwen, de Wit & Smit, 2017); only one resident included references to LPG in their statements.

Energy usage

Considering the significance of electricity in the last section it is no surprise that the demand for electricity will be a large part of the energy usage cultural domain. Electricity demand is a complex system of factors and each factor could itself be thought of as a system of like complexity (Buys et al., 2015), for example appliance usage, household demographics, culture, weather and climate, and retail electricity rates (Engle, Mustafa & Rice, 1992; Hong, Chang & Lin, 2013; Buys et al., 2015).

This demand varies between and within any given day or season, and is made up of a baseload, that is, a demand level that is always present, and a peak demand, that is, an increased demand that results from higher consumption based on society's collective energy activities (Rallapalli & Ghosh, 2012; Dahlke & McFarlane, 2014/2015). The latter is a sign of societal synchronization (Torriti, 2017) and is made up of two primary factors: the amount of energy and the timing of when this energy is used (Yarbrough et al., 2015).

Baseload

Residential energy consumption is not as tangible as other forms of consumption (Buys et al., 2015) especially in modern energy systems. Sovacool (2009) and Mazur-Stommen (2013) describe the physical invisibility of energy technologies and energy use, but Sovacool suggests that technologies that are highly visible are often believed to be more consumptive by its users. This is partly because energy in homes is connected to energy-using appliances and no longer directly to the fuels that power them (Rüdiger, 2008).

The residents interviewed acknowledge that specific appliances, that is material culture have substantial impacts on energy consumption, for example:

> The appliances that use a lot of energy like the air conditioner, hot iron for ironing clothes and [the] washer-dryer – so, any heating or cooling appliances.
>
> (Resident 10)

Quotes like these also include norms, practices and external influences. But they are particularly useful to show how observable the material culture references are; in this example the material culture is the air conditioner, hot iron and washer/dryer.

The best way to present norms and practices would be to discuss both together but to make the distinction between them clear. From the quote provided earlier, an example of a norm is the belief that switching on and using (the practice) heating and cooling appliances (the material culture) will increase the amount of energy used in the home. Another example is:

> If it's [the appliance] not running all the time then the bill would not go up.
>
> (Resident 1)

As hinted by this quote, the length of time an appliance is used is a practice. However, by using appliances (the material culture and practice) for longer periods of time, the resident believes and expects that more energy will be used, and this is the norm. This example plays on the fundamental technical relationship between the power rating of an appliance and its energy consumption where the length of time it is used for is a key variable factor.

Most of the residents provided responses like those presented thus far. But there is a general question arising as to why material culture is used in the way that it is. Looking at the householders' external influences helps answer this question. The external influences in this domain were the tropical heat and the utility company's influence.

There are environmental influences on the island and residents reported the main factor to be the tropical heat and the subsequent need for cooling, for example:

> The heat – we have 4 air conditioners which are being used much more than expected since they're used day and night.
>
> (Resident 16)

This illustrates that weather directly affects thermal load and energy use (Hong, Chang & Lin, 2013), and that the prime driver of cooling demand is the climate (DeForest et al., 2014). As shown by the quote above, the tropical heat is an external influence which causes householders to switch on (the practice) their air conditioners (the material culture) during the day and night, and by using the air conditioners in this way, there is an aspiration to cool the home and the expectation that it will be cooler (the norms).

In terms of the utility's influence, cost recovery, reliability and electricity costs were raised. Relative to the electricity costs, one resident simply stated that the costs influenced their energy usage. But relative to the utility's reliability, another thought that good quality service is not always provided by utilities. This resident also stated that earlier that week (prior to the interview) there were several power cuts to their home over a short period of time and went on to say that those sorts of occurrences could damage appliances in which case they asserted that they are not compensated for any damages that may be incurred. As it pertains to the utility's cost recovery, this same resident also thought that the company's expenses are frankly passed on to consumers.

Peak demand

The preceding section gives a good descriptive overview of the baseload dynamics of 'energy usage' as a cultural domain in the mainstream energy culture sampled. However, a dimension that is missing is that of time – a key element of energy services (Sorrell & Dimitropoulos, 2008). In that regard, Figure 5.2 shows the number of residents who selected respective hour-long timeslots which they believe represent their household's perceived peak energy demand times.

Electricity and mainstream energy cultures 101

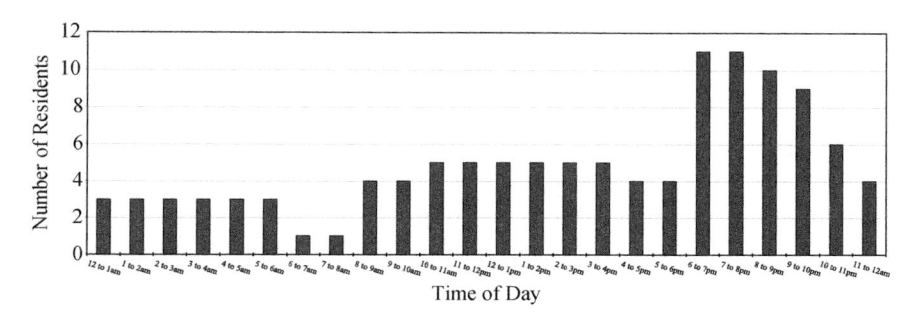

Figure 5.2 Showing the perceived peak energy demand times of the residents interviewed in Trinidad.

The graph shows that there are variations in when householders think the most energy is used in their respective homes. This variation is likely due to lifestyle diversity, family makeup as well as the residents being able to answer the question using what they personally believed to be a 'typical' day in their household, that is, weekday versus weekend perceptions. Nevertheless, the explicit practices driving these time selections include examples such as use of lighting, televisions, laundry/washing, cooking and showering, for instance. A caveat is that many residents simply stated that their selections were based on generic energy uses that were not specific, for example the need for cooling at the hottest times of day. However, the data led to four central takeaways.

First, there is an acknowledgement that energy is constantly being used in the home by various appliances whether they technically use power all day, for example refrigerators and deep freezers, or people think they use these appliances enough to warrant them believing so, for example televisions, air conditioners and computers – hence another reason for the erratic data in Figure 5.2.

Second is that there are energy activities happening in the home throughout the day. These are a mix of active and passive practices. Examples of active practices are cooking, washing and showering. The passive practices largely refer to the latent ones that can be deduced from the continuous use of material culture – though latent practices can also be active. Both show that the design and use of domestic appliances influence the sequence and duration of activities and the timing of demand (Carlsson-Hyslop et al., 2016). But the fundamental difference between active and passive practices is the human presence in the home to perform the practices.

An example of a latent, passive practice is refrigerator usage. Simply leaving it plugged in means that its services are used independently of the household being present in the home; they still use and benefit from it storing and chilling their produce/goods. A latent, active practice would again draw on material culture as a proxy and television usage is an example where it is switched on to view aired programmes. A manifest, active practice example is coming home after the day's work to cook dinner. Many different appliances, for example electric/gas stove, blender and/or microwave oven would be used to prepare the meal and so cooking requires direct and sustained use of material culture to perform the practice.

The third point is the most visible in Figure 5.2. There is consensus relative to when the residents' perceived peak energy times occur. The broad peak occurs between 6:00pm to 11:00pm. The more specific peak times observed, 6:00pm to 8:00pm, were selected by 11 residents for each hour, that is, 6:00pm to 7:00pm and 7:00pm to 8:00pm. These can also be compared to the wider grid's technical peak demand timing that occurs roughly between the same times (Marzolf, Cañeque & Loy, 2015, 269–271).

The residents' sense of time is not tied to monitoring timepieces but rather to their immersion in routines and habits. Habits and routines are observable performances and the patterning of stable practices which are recurrent, non-reflexive and culturally shared (Gram-Hassen, 2008; Southerton, 2012). Routines sustain everyday life as well as its wider socio-cultural and technological structures which means that routines have an inertia (Gram-Hassen, 2008).

Habits are defined by their frequency of repetition and are somewhat automatic because past and future actions are linked and suggest that people maintain traditions (Aarts & Dijksterhuis, 2000, 53). Present practices are therefore predisposed to continue in the future (Wilhite, 2013; Wallenborn & Wilhite, 2014), and they may be likely repeated at the same times on any given day as well (Torriti, 2017).

The perceived peak demand times show that the material, infrastructural and institutional environments can make practices recurrent and non-reflexive (Gram-Hassen, 2008; Southerton, 2012); they support the materialization of energy through the way in which individuals use and access matter as routine undertakings (Sheller, 2014), and make energy a daily part of life (Muhammad-Sukki et al., 2014).

The fourth and final point is that the perceived peak demand timings are largely based on daily lifestyles despite some sensitivity to weekly and even annual timeframes. These timeframes are a result of external institutions' influences on people's presences in their homes, for example schools and workplaces. This shows that the demand rhythm is not solely an in-house product (Carlsson-Hyslop et al., 2016) and external institutions are also an example of external influences in the ECF and the landscape in the MLP.

There are everyday practices that are grouped together based on their daily importance in people's lives and these practices are created, sustained and transformed through everyday life (Strengers, 2010). Therefore, this notion of 'daily importance' means that there are material culture, norms, practices and external influences which are more salient than others relative to how energy is used in homes.

So, to put the earlier-outlined peak demand insights into more explicit context relative to the baseload's constructs, a rudimentary free-listing exercise was done. This provides a list of features that are ranked according to their socio-cultural salience. There were 36 different explicit 'items' recorded. Of these, roughly 88% are appliances, for example refrigerators, and the remaining are a mix of practices, for example cooking and showering. This means that the residents' material culture are the most salient features driving the householders' perceived peak energy demand.

The REC (2011; 41) provides information on an average Trinidadian home's consumption breakdown and it was compared with the appliances' saliences described above, that is, a comparison between their technical consumption and cultural salience. The most consumptive material cultures are not necessarily the most socio-culturally salient, for example electric water heaters. In the case of electric water heaters, this is likely because householders have skewed perceptions of their energy demands and underestimate the energy consumption from heating water (Sovacool, 2009; Lockton et al., 2013).

But an interesting point to flag here relative to the residential energy transition through solar energy is that if hot water is not culturally salient in Trinidad as inferred by the salience of electric water heating, then technologically transitioning to SWHs would not be in householders' direct line of sight regardless of the potential to reduce their electricity bills and/or consumption. Therefore, increasing the visibility of socially invisible services is paramount for sustainable end-user behaviour (Pink, 2011; Anda & Temmen, 2014).

Further, the material culture are appliances that residents owned. So, when put into context of this mainstream energy culture sample, it shows that the appliances which make up the material culture are those that are owned versus residents having knowledge that these appliances can impact their energy usage even though they are not owned/used in the household.

Other data that arose organically indicate that there were in-house technological transitions. Statements converged on a lighting transition from incandescent lighting to either a mix of fluorescent and incandescent or all fluorescent. This observation showed that homes experience technological transitions which provide the same service (in this case lighting) but through different technologies (incandescent versus LED or fluorescent lighting). These thoughts therefore invite considerations for the potential technological transitions from the incumbent forms of water heating to SWHs and how that could play out given the data from Table 5.2.

Hot water use and the transition to solar water heating

Table 5.2 shows the hot water material culture used by the residents interviewed and what they are powered by.

Electric water heating is the most significant method (see Table 5.2); technologies used include storage tank, showerhead point-of-use and wall-mounted

104 *Starting the household solar transition*

Table 5.2 Showing the water heating material culture of the residents interviewed in Trinidad

Frequency	Material culture	Powered-by
5	Storage-tank electric water heater	Electricity
3	Showerhead point-of-use electric water heater	
1	Wall-mounted point-of-use electric water heater	
2	Electric kettle	
1	Microwave	
1	Electric stove	
3	Gas stove	LPG
1	SWH	Solar energy

Source: The author.

Note
The frequency is defined as the number of responses that are tagged with the associated material culture.

point-of-use water heaters as well as kettles, microwaves and stoves. Three residents also indicated that they combusted LPG through their stove burners to heat water (see Figure 6.2) – even the resident who reported using the electric stove (see Table 5.2) stated that it used LPG as well. Interestingly, one resident owned an SWH, and this was an unexpected observation during their interview.

Nevertheless, based on the last section's ending discussion, it seems that a transition towards SWHs in Trinidad will involve the substitution of electricity and LPG-based water heating for solar energy. The emergence of these two resources also confirms the key finding from the chapter's first section: electricity and LPG are key household energy inputs. But understanding what hot water is used for (if at all) will help build a better understanding of the transition to SWHs. This is useful because material culture like an electric kettle can be used to heat water for making tea or even bathing, for instance, which are different practices.

There is a limited suite of hot water practices which relate mainly to showering/bathing, preparing tea, cooking and washing dishes. One resident however, added that all their pipes have hot water so in this case there is a greater potential for them to use it more ways, for example washing hands and cleaning. These observations also confirm the earlier reported findings which show that hot water use is part of the residents' everyday routines. But to bring this hot water narrative full circle relative to its immersion in everyday routines, it is worth considering the reasons why water heating would have potentially low cultural visibility. There are three reasons.

Householders think that bigger appliances often consume more energy (Steg, Perlaviciute & van der Werff, 2015). So, the first reason is that some of the water heating appliances used are rather small, for example showerhead point-of-use electric water heaters. This is also coupled by the fact that the larger components of some appliances may be installed in a location that makes it invisible in householders' daily lifestyles, for example the storage tank for some electric water heaters.

A second reason is that electricity is not exclusively used for heating water nor is combusting LPG. The material culture that use electricity are numerous and heating water is just one energy service amongst the others. So, it is easy for water heating to be culturally lost energy-wise. Relative to LPG use, heating water is often not the main activity, for example combusting LPG for cooking where heating water may be just part of the meal preparation.

The third reason is that there may be no actual or perceived hot water usage. Two residents said that they did not use hot water and their hot water use was minimal:

> What I have is what I need.
> (Resident 12)

> There's no real hot water use.
> (Resident 14)

Resident 12's wider response implied that hot water is a luxury and not a necessity. In Resident 14's case, they used to own an electric water heater but with today's perceived high temperatures, they thought that taking cold showers is a comfort hence why they believed that their hot water usage was insignificant.

Energy costs

The energy costs domain was explored in a similar fashion to the energy usage one and residents pointed out norms, material culture, practices and external influences that they believe affect the costs of the energy their household uses. Again, the term 'energy' was used over 'electricity' to keep the questioning consistent with the previous domains, to let any electricity-related responses emerge organically, and to ensure that there was still scope for the inclusion of other non-electrical factors.

As with the baseload described earlier, the cost domain does not have an explicit consideration for time. However, as will be shown, the earlier peak demand dimension puts some of the features that will be discussed into the context of the residents' routines – which is useful because this domain was defined by energy consumption and its retail electricity costs as broad themes/concepts.

As noted with the baseload from before, air conditioners, televisions, lights and washing machines are examples of material culture, for example:

> Washer; dryer; refrigerator; lights at night.
> (Resident 7)

Additionally, there were similar norms and practices, for example those related to the length of time an appliance is used for based on statements such as:

> frequency of use of the different energy-consuming things in the house.
>
> (Resident 16)

In this example, the resident is saying that by using more appliances (the practice and material culture), they believe and expect their energy costs will increase (the norm). There were no significant differences in the types of material culture, norms or practices and this seems to suggest that the energy usage and costs domains overlap markedly. But there were some differences in external influences; they are largely environmental and institutional-industrial.

An example of environmental influences comes from one resident stating that their air conditioning might be used more frequently during the dry season whilst their clothes dryer may be used more frequently in the rainy season. The dry season extends from January to May (Narinesingh et al., 2013) and is the drier part of the year with higher temperatures which therefore create increased cooling demands. On the other hand, the rainy season's (June to December) impact on drying clothing was presumably due to less sunshine (due to more cloud cover) that would have otherwise been used for drying clothing outdoors in addition to possible drenching by rainfall.

With respect to the other external influences, they are associated with the retail electricity costs. This is driven by industrial costs and the electricity bill – where the latter is the most significant means of communication between electricity suppliers and consumers because of its cost and consumption information (Winther & Bouly de Lesdain, 2013), and is also one of the only correspondences between both parties (Morris, Vine & Buys, 2016).

Other influences are the fuel and fixed costs, the retail tariff itself, local authorities and their cost recoveries, as well as the utility's billing methodology. For example:

> the relevant mark-ups in the line agencies, [the] T&TEC and [the] Power-Gen, in providing the service.
>
> (Resident 15)

> we don't have control over the rates since they are fixed prices.
>
> (Resident 16)

The utility's billing methodology is worth outlining further since it begins to take the discussion directly into the residents' quantitative costs of energy. One resident believed that the utility uses the average consumption for a given area in the island to determine individual households' electricity bills and does not necessarily use homes' actual electricity meter measurements;

the meter measurements are rated at US$0.04 for the first 400 kWh, $0.05 per kWh for the next 600 kWh, and anything over 1,000 kWh is charged at $0.06 per kWh (Espinasa & Humpert, 2016).

This resident continued by stating that after speaking to friends living in urban areas, they found that their friends' bills were near US$330, others in the more rural areas were closer to $70, whilst theirs was closer to $180. They acknowledged that appliance usage can account for such disparities but believed that their urban friends tried reducing their consumption, yet the bills were perceivably high. However, they were unable to give any figure for the cost of the bills after this consumption reduction experiment. But the next section puts these sorts of accounts into the context of bills being either too high, affordable or too low by describing the drivers of these perceptions.

Perceived electricity bills

The residents interviewed provided estimates of an 'average' electricity bill for their respective households. This was important because having some cognitive quantities to consider in tandem with the factors which affect households' energy costs (as well as their usage) helped to better frame their accounts. Residents' bills ranged from less than US$42 to $183.

But to give this spread meaning, the residents further gave their opinions on their bills. So doing gave some qualitative sense of what these costs mean to people. The responses largely fit into answers that could be described as affordable and reasonable cost perceptions (i.e. the costs of electricity are not too high, are reasonable, or are more affordable than before); high cost perceptions (i.e. the costs of electricity are too high or are too high sometimes); or being simply unsure. There are four key observations worth pointing out.

The first is that most of the responses revolved around the fact that the costs of electricity are either too high or not – which is likely because Trinidad's retail tariffs are non-fluctuant (unlike the other case studies). So, there can be an established cognitive baseline relative to residents' 'average' bills. The two variables in these bills are household energy consumption and the retail rate; in Trinidad constant consumption levels over subsequent months would produce a similar bill but in the other case studies, however, the fluctuant global oil price and consequent retail rate means that even though household consumption may remain constant, it does not mean that the bill will be.

Second, at the individual level, sometimes a person's response acknowledged more than one reason driving their perception(s). For example, one resident who thought that their bill was too high believed so because they compared their electricity bills to their other utility bills, but also because of the type of oil that they believed was used for generating power (though oil is not used for power generation).

The third observation is that the drivers of perceptions can be interpreted differently by individuals such that there are opposing views emerging from the same reason(s). For example, the perceived geographic billing designations illustrated this. Several residents thought that electricity billing was done based on location (not metering) and urban households would generally have higher bills than rural homes; one resident who lives in a rural area viewed their bill as not too high whilst another who lives in an urban area thought it was too high.

Fourth, unlike the other energy culture dimensions investigated, there was no overwhelming consensus during the analysis. This is likely the result of the highly personal nature of the perceptions playing out in the interpretation of electricity costs by householders. Singular opinions predominantly drove the perceptions of electricity costs as being high except for two residents who believed the costs are high due to the generic cost of living, for example:

> Relative to the cost of living, the rate per unit of electricity i.e. kilowatt-hour is too high.
>
> (Resident 10)

Two others also viewed the country's perceived wealth and its local natural gas resources as reasons that electricity should be a free commodity. So, they believed that the current costs are too high, for example:

> We have natural gas; that's (i.e. electricity) supposed to be free.
>
> (Resident 11)

Affordable/reasonable price drivers revolved around the belief that the residents' energy costs were commensurate with their usage – for instance:

> because it [i.e. electricity] is used all the time.
>
> (Resident 1)

The costs of electricity were also perceived as affordable/reasonable by several residents because in comparison to their peers' bills, theirs were cheaper.

Solar energy and solar water heating

The next chapter focuses on the transition to/through solar water heating. So, to give a sense of continuity, now Trinidad's mainstream energy culture will be described a bit more explicitly along those lines. The residents were asked about solar energy and solar water heating. Their responses provide a 'baseline' for householders' perceptions of solar energy as a resource and solar water heating as a technology in a place which has not undergone the solar energy transition.

The interviews therefore distinguish between the resource and technology so that each could be studied in relation to Rogers' innovation-decision process, that is, the process potential adopters go through when making an adoption decision: knowledge, persuasion, decision, implementation and confirmation (see Rogers, 1983, and the mainstreaming theory from Chapter 2).

Both dimensions in this chapter, that is, solar energy and solar water heating, focus on the knowledge stage of this process. These are useful dimensions to look at in Trinidad because the responses would have shown what solar energy norms exist, if any, independently of adopting the technology.

The meaning of solar energy

Analysing the residents' responses related to solar energy led to three central observations.

First, the fundamental acknowledgement that the sun is the energy source and sunlight is the resource was explicitly highlighted by 11 householders through responses such as:

> Energy from the sunlight.
>
> (Resident 7)

This makes the acknowledgement of the sun being the energy source the most salient aspect.

Additionally, two residents suggested that solar energy is a form of natural energy; one described it as being made up of different wavelengths, and another acknowledged its extra-terrestrial origins as shown by the following:

> Harnessing the natural energy that comes from the sunlight – [and] filtering it to power households or anything.
>
> (Resident 5)

> Energy that comes directly from the solar system.
>
> (Resident 13)

> The energy transmitted by all the various wavelengths from the sun.
>
> (Resident 15)

Second is the use of technology to make solar energy practical; three responses mentioned the need for a solar collector surface, for example:

> There is a glass dish [a solar collector surface] that is put out in the sun to store current.
>
> (Resident 9)

> Solar energy is using the sunlight [by] using solar panels that convert the electricity and store it in batteries.
>
> (Resident 14)

> Provision of power through cells which are themselves receiving energy from the sun.
>
> (Resident 16)

An interesting observation from these quotes is that the technological references are largely related to PV which convert the sun's radiation into electricity that is then transformed by an inverter to the appropriate currents and voltages (van Leeuwen et al., 2017). This is not surprising since the residential market is the most attractive market for PV (Islam, 2013), and PV panels have come to epitomize rooftop solar as the mainstream solar technology globally (WEF, 2013; WEC, 2016).

Further, four residents' responses suggested that using solar energy involved an energy conversion. Resident 14's earlier quote is an example, and another is:

> and has a plug that attaches it to a converter as well as a cable which connects it to the household so that the various household appliances can be used.
>
> (Resident 9)

This response particularly suggests that a converter is required to use solar energy. PV systems produce a DC. However, the utility's electricity provided to homes is an AC. So, depending on the type of appliance either DC or AC can be used. But when the latter is needed from PV, an inverter is required to convert DC to AC (Goss, 2010; The NEED Project, 2016); the converter being referred to by Resident 9 was more than likely this inverter.

Other technological references are energy storage and wiring. Resident 9 and Resident 14's quotes show the only storage-related answers given by the residents interviewed. Resident 9's quote identifies with the cabling needed to connect the solar system to the home's energy appliances and was the only response that referred to wiring.

The third and final observation is related to solar energy's relative advantages. Relative advantages refer to the extent to which an innovation is perceived as being better than the one it supersedes, and it is often expressed through economic profit and social status (Rogers, 1983). So, this means that solar energy is perceived to have advantages over the incumbent household energy used, that is, the electricity and LPG outlined in the chapter's first section.

At the systems-level, common examples are reduction in fossil fuel consumption; need for constructing conventional power generation stations; greenhouse gas emissions; health implications from emissions; blackouts and power outages; peak energy demand; electricity transmission and distribution costs; unemployment; and dependencies on finite resources (Grover, 2007; Naspolini, Militão & Rüther, 2010; Kabir, Kim & Szulejko, 2017; Niu et al., 2017; Kabir et al., 2018).

At the building level, common examples are increased property value; localized electricity generation; use of a freely available resource; economic benefits from incentives (if available); low maintenance costs; convenience in terms of time and effort spent on sourcing and paying for energy in some cases; reduced energy bills and consumption; lower dependence on public infrastructure; and low operational environmental impact (Naspolini, Militão & Rüther, 2010; Leaman, 2015; Kabir, Kim & Szulejko, 2017; Niu et al., 2017).

In the case of the interviews, there are three responses that connected with solar energy technologies' potential to save on energy costs, for example:

> Using the resources of the sun to save money.
>
> (Resident 6)

> Something that could cost you less than [the] T&TEC.
>
> (Resident 8)

Resident 6's quote hints that because sunlight is freely available versus paying for the incumbent energy used, the sun's energy is directly translatable into monetary savings. Resident 8's quote shows that there is the belief that the T&TEC's services cannot only be substituted with solar energy, but also that solar energy can act as a substitute for the present sources of electricity.

But these came with one resident acknowledging that the costs of solar energy are still somewhat of a moderating factor at present – given that solar technologies usually have high initial investment costs (Naspolini, Militão & Rüther, 2010; Mathews & Mathews, 2016; Kabir, Kim & Szulejko, 2017).

One other resident suggested that another advantage would be appliance damage mitigation where any appliance damage due to power outages could be reduced if solar energy is used. Additionally, Resident 11 implied that solar energy is further advantageous because hydrocarbons are finite when they said that:

> at some point in time, oil and natural gas will become limited.
>
> (Resident 11)

A final relative advantage was mentioned by Resident 3 who suggested that solar energy is a strategy that can mitigate climate change since:

> it helps with addressing global warming.
>
> (Resident 3)

So overall, this snippet of the responses provided by residents suggest that their norms tied to 'solar energy' are centred on it as an energy (re)source but are also sensitive to its technologies and the technicalities behind using it, as well as its relative advantages. But how would these solar energy norms relate to solar energy technologies? The answer was found by analysing the norms related to solar water heating.

The meaning of solar water heating

In the interest of conciseness, the central conceptualization of how the residents believed a SWH works will be focused on through a five-part functional narrative.

First, residents acknowledge the sun as the source of energy for solar water heating and that it is only available during the day, for example:

> Heating the water using the sun.
> (Resident 1)

However, this is sensitive to the fact that sunlight is made up of different wavelengths and SWHs are believed to capture a certain wavelength of the spectrum, for example:

> Converting the relevant [light] spectrum from the sun to provide energy for [the] warming up of the water for your shower.
> (Resident 15)

There is also the belief that a SWH can function during rainy periods because it is able to utilize the ambient heat beyond direct sunlight according to one resident.

Second is that the incoming solar energy is believed to be captured by a solar collector that coverts solar radiation into heat and there were references to collector surfaces, panels, cells and glass dishes, for example:

> There are also solar water heater tanks with solar collectors used to heat and store water.
> (Resident 2)

> Solar panels trap sunlight and has the ability to store this energy as an attachment to a water heater to transmit the energy to heat the water.
> (Resident 13)

Third, residents suggested that a solar collector enables an energy conversion where there is a heat energy transfer such that thermal energy is trapped, stored and even reflected during the actual water heating. The heat is also thought to be applied to the water via circulation plumbing.

For example, Resident 6 thought that the system has metal pipes which enable the water heating as part of the circulation system. Another, Resident 14, said that some solar water heating technologies use coils to heat the water. Nevertheless, the hot water plumbing leads to a storage tank. Responses include quotes such as:

> There are metal tubes made out of a material that stores heat – [they are] surrounded by glass that gets hot. The top of the tubes run into the

Electricity and mainstream energy cultures 113

storage tank and the water is transferred down [from the storage tank] into the pipes.

(Resident 6)

I never saw one in real but through the internet, [specifically] YouTube, there seems to be water pipes involved where some do-it-at-home versions have foil paper that reflect heat towards the piping and others that use coils for heating.

(Resident 14)

Fourth, two residents outlined that SWHs use a storage tank that is connected to a water circulation system. Interestingly, one believed that solar water heating was different to electric water heating because the former cannot heat running water which meant that it needs a storage reservoir to be effective – showing how the earlier described electrical water heating transition to SWHs is subtly intertwined.

The fifth and last element is residents pointing out that SWHs are a type of technology which needs to be installed. Resident 2's answer was of particular interest because they outlined that solar water heating can be installed on the ground or roof of a home. Though they did not mention a specific SWH design, their views surrounding installation drew on references to passive systems, that is, systems which use gravity for water circulation:

There are several different types. Piping from natural sunlight where one lays a grid of PVC [Polyvinyl Chloride] pipes on the roof and the sun heats it – it gets very hot. Laying it on the roof allows the water to flow. Pipes can also be in the ground and that gets hot as well.

(Resident 2)

With this idea of the norms related to SWHs, the overall idea of the system could best be summarized as a technological installation that converts sunlight into thermal energy which is transferred to the water circulating in the system and later stored for providing hot water.

But beyond this general understanding there are also potential technological misconceptions which are worth highlighting. There were references to PV which is solar electricity versus the solar thermal used for heating water (not discounting the fact that PV can be used to heat water). Residents outlined PV systems' connections and the potential for energy storage through responses such as:

There is a glass dish [a solar collector surface] that is put out in the sun to store current and has a plug that attaches it to a converter as well as a cable which connects it to the household so that the various household appliances can be used.

(Resident 9)

These PV references are also accompanied by references to smaller-scaled applications such as solar lighting, for example:

> [Solar water heating] Works on a similar principle as the solar garden lights.
>
> (Resident 1)

> I have solar lighting outside, but I only bought 1 as a trial before buying more. It has worked consistently well so far so solar water heating should work well also.
>
> (Resident 16)

Further, these misconceptions come with some persons simply not knowing how a SWH worked, having never seen the technology, or even not being aware it exists.

All these knowledge-related insights show that residents have some awareness of the technology in question, that is, the SWH, but the latter misconceptions indicate that there are information gaps that would have to be filled before a household adopts the technology – and Resident 7 summarized this aptly when they said that:

> education is important since people's knowledge is limited.
>
> (Resident 7)

The interviews gave the impression that the residents were interested in gathering more information about SWHs as a technology before considering adopting one – which relates to the knowledge phase of the innovation-decision process (see Rogers, 1983). The residents accounts also suggest that they would use a range of media to gather such information, for example:

> Since I like technical things, the internet through sites like YouTube may be best since it has a lot of resources.
>
> (Resident 14)

> I would visit a supplier or consult with someone that has expertise like an engineer or someone who has experience with owning solar.
>
> (Resident 4)

> call up a friend that has had dealings with solar water heating in Barbados.
>
> (Resident 2)

> the media should also help inform the people of the advantages of solar over electricity, not to put the T&TEC out of business, [but rather] so that people can have a choice; awareness is the key.
>
> (Respondent 3)

Resident 2's quote is significant for two reasons: first, it shows that social networks are not limited to intra-island locals, but cultural connections exist between islands, and second, it specifically includes Barbados, the case study for the cultural impact of solar water heating.

Summary

Electricity and LPG are the salient household energy (re)sources in the mainstream energy culture sample from Trinidad. However, the technical significance of electricity raised in Chapter 4 has been translated culturally as well – residents are even aware that it is generated from fossil fuels. Chapter 4's regime-wide dynamics are culturally translated because a home's infrastructure is a cultural energy space for its residents.

But at the interface of technical power-provisioning and the energy culture, 'electricity' is the culturally salient norm. A cultural lock-in exists such that items of material culture are largely household appliances that have been designed to run on electricity, and many of the norms and practices have been institutionally locked-in to the relationships between homes and the T&TEC. This demonstrates that electricity as a cultural notion is part of the regime but also the landscape of the MLP from Chapter 2.

The energy usage and costs cultural domains heavily overlap because they share material culture, norms, practices and external influences. The energy usage domain, however, can be explored as the baseload, that is, the demand for energy that is always present, as well as the peak demand, that is, the increased consumption that results from society's synchronized energy use at specified times. There is a degree of consensus on the fact that the perceived peak energy demand occurs in the evening/night, and this is the same technically as well.

This is largely because the popular notion of a 'typical' day relates to weekdays. It means that the energy culture is strongly affected by the external influence of institutions such as workplaces and schools; the most energy is used during evenings/nights because householders are more likely to be at home then. These considerations therefore put energy usage into the context of residents' lifestyles, routines and habits which are daily phenomena that are repeated enough to scale-up into a longer-term temporal feature in the MLP's landscape.

Relative to energy costs, there is a marked variation in the perceived costs of electricity by householders based on the reported 'average' bills. This is likely due to the highly personal nature of electricity consumption. Still, there is some rudimentary polarization of perceptions such that bills are perceived as either being too high and either too low/affordable which is likely due to the fact that the retail electricity rates in Trinidad do not fluctuate with the international oil prices unlike in Barbados and Oʻahu.

Given the cultural lock-in brought about by electricity-provisioning, consumption and paying bills, people seem to connect with the fact that using more energy incurs higher costs – which is independent of changes in electricity prices. Further, people's use of their material culture helps explain the relationships between energy usage and its costs, for example using more lights (the material culture) means higher energy consumption and costs.

Material culture is arguably the most common way people identify with and explain their energy culture. It is also a cultural proxy for energy services and the associated norms, practices and external influences. With the above lighting example, for instance, lights are the material culture; a norm can be the expected increased visibility due to the illumination; flipping the switch is the practice; and the darkness of nightfall can be an external influence.

Related to energy services and households' material culture are the technological transitions happening in homes – and transitioning to options like SWHs may be one. The lighting transition reported from incandescent to LED and/or fluorescent options illustrate that the energy service, that is, lighting, remained constant though the technology changed.

So, the heating service through hot water may be constant despite there being a change to SWHs from incumbent options like electric water heating and/or LPG as would be the case in Trinidad. What is interesting to mention here is that water heating is not the most culturally salient use of energy in households. So, this will have implications for adopting SWHs because if the service and elements of the incumbent technologies are not as visible, then transitioning to SWHs would not readily occur even though doing so can save money and energy.

Nevertheless, an energy culture has material culture, norms, practices and external influences which have varying socio-cultural saliences over time. Therefore, there is arguably no fixed constellation of such features in an energy culture or at any one point in time in an energy culture. Changes in the salience of these features are time-dependent and technological transitions are an example of a process that may spur such changes.

But despite not having assimilated solar energy technologies like SWHs, the residents' mainstream energy culture still includes norms associated with solar energy. There are norms relating to solar energy being from the sun, requiring technology to be usable, and having relative advantages over the incumbent electricity supply. In the case of the norms related to SWHs, they are based on the beliefs and expectations that they are technologies which convert the sun's radiation into thermal energy to heat water circulating in the system and which is then stored in a storage tank to be used when needed. Further, these perceptions are riddled with PV references, so this suggests that PV is the more culturally familiar of the two technologies based on what residents believe and expect solar energy and solar water heating to be.

Mainstream energy cultures are practical and 'lived-in'. They vary in structural complexity from one regime to another and the cultural domains linked to the energy usage, costs and supply of the energy culture have multiple features that can be more strongly expressed in one cultural domain than another and/or even in one energy culture than another. This demonstrates that energy cultures are highly convoluted and shows why there is no distinct boundary between what constitutes 'culture' in the MLP.

However, the question to ask now is: how does a mainstream energy culture which has adopted SWHs differ to the one described in this chapter? To answer this, Chapter 6 looks at the second case study, Barbados.

References

Aarts, H. & Dijksterhuis, A. (2000). Habits and Knowledge Structures: Automaticity as Goal-Directed Behaviour. *Journal of Personality and Social Psychology*, 78(1): 53–63. Retrieved from https://psycnet.apa.org/doi/10.1037/0022-3514.78.1.53.

Anda, M. & Temmen, J. (2014). Smart Metering for Residential Energy Efficiency: The Use of Community-based Social Marketing for Behavioural Change and Smart Grid Introduction. *Renewable Energy*, 67: 119–127. Retrieved from https://doi.org/10.1016/j.renene.2013.11.020.

Broesch, J. & Hadley, C. (2012). Putting Culture Back into Acculturation: Identifying and Overcoming Gaps in the Definition and Measurement of Acculturation. *The Social Science Journal*, 49(3): 375–385. Retrieved from https://doi.org/10.1016/j.soscij.2012.02.004.

Buys, L., Vine, D., Ledwich, G., Bell, J., Mengersen, K., Morris P. & Lewis, J. (2015). A Framework for Understanding and Generating Integrated Solutions for Residential Peak Energy Demand. *PLoS ONE*, 10(3): e0121195. Retrieved from https://dx.doi.org/10.1371%2Fjournal.pone.0121195.

Carlsson-Hyslop, A., Kuijer, L., Shove, E., Spurling, N., Trentmann, F. & Watson, M. (2016). *The Timing of Domestic Energy Demand. Insights form the 1920s – 2000s.* DEMAND Centre. Retrieved from www.demand.ac.uk/wp-content/uploads/2016/10/DEMAND-insight-8.pdf.

Citizens Advice. (2016). *Tackling Tariff Design Making Distribution Network Costs Work for Consumers.* London. Retrieved from www.citizensadvice.org.uk/about-us/policy/policy-research-topics/energy-policy-research-and-consultation-responses/energy-policy-research/tackling-tariff-design-the-tariff-transition/.

Dahlke, S. & McFarlane, D. (2014/2015). *Demand Response: Reducing Peak Electricity Demand.* Great Plains Institute, Minneapolis, Minnesota.

DeForest, N., G. Mendes, M. Stadler, W. Feng, J. Lai & C. Marnay. 2014. *Thermal Energy Storage for Electricity Peak-demand Mitigation: A Solution in Developing and Developed World Alike.* Ernest Orlando Lawrence Berkley National Laboratory. Presented at the 2013 European Council for an Energy Efficient Economy, Belambra, Les Criques, France, 3–8 June 2013. Retrieved from http://escholarship.org/uc/item/0nx5q3dq#page-1 (accessed 4 March, 2017).

Demirel, Y. (2016). Energy and Energy Types. In *Energy, Green Energy and Technology* (pp. 27–70). London: Springer International Publishing. Retrieved from https://doi.org/10.1007/978-1-4471-2372-9.

Engle, R. F., Mustafa, C. & Rice. J. (1992). Modelling Peak Electricity Demand. *Journal of Forecasting*, 11(3): 241–251. Retrieved from https://doi.org/10.1002/for.3980110306.

Espinasa, R. & Humpert, M. (2016). *Energy Dossier: Trinidad and Tobago.* Inter-American Development Bank. Retrieved from https://publications.iadb.org/en/energy-dossier-trinidad-and-tobago.

Fryberg, S. & Rhys, R. (2009). *Cultural Models*. Retrieved from www.scribd.com/document/296580946/Fryberg-Cultural-models.

Goss, B. (2010). *Choosing Solar Electricity: A Guide to Photovoltaic Systems*. Machynlleth, UK: Centre for Alternative Technology.

Gram-Hassen, K. (2008). *Energy Consumption in Homes – An Historical Approach to Understanding New Routines*. In M. Rüdiger (Ed.), *The Culture of Energy* (pp. 180–199). Newcastle: Cambridge Scholars Publishing.

Grover, S. (2007). *Energy, Economic, and Environmental Benefits of the Solar America Initiative*. National Renewable Energy Laboratory, Colorado. Retrieved from www.nrel.gov/docs/fy07osti/41998.pdf.

Hiemstra-van der Horst, G. & Hovorka, J. (2008). Reassessing the 'Energy Ladder': Household Energy Use in Maun, Botswana. *Energy Policy*, *36*(9): 3333–3344. Retrieved from https://doi.org/10.1016/j.enpol.2008.05.006.

Hong, T., W. Chang & H. Lin (2013). *A Fresh Look at Weather Impact on Peak Electricity Demand and Energy Use of Buildings Using 30-Year Actual Weather Data*. Ernest Orlando Lawrence Berkley National Laboratory, California. Retrieved from http://escholarship.org/uc/item/3174x75w#page-1.

Horta, A., Wilhite, H., Schmidt, L. & Baritaux, F. (2014). Socio-technical and Cultural Approaches to Energy Consumption. *An Introduction: Nature and Culture*, *9*(2): 115–121. Retrieved from https://doi.org/10.3167/nc.2014.090201.

Islam, T. (2013). *Stated Innovation Diffusion Model from Stated Preference Data: The Case of Photo-Voltaic (PV) Solar Cells for Household Electricity Generation*. Paper presented at the International Choice Modelling Conference, Sydney, Australia. Retrieved from www.semanticscholar.org/paper/Stated-innovation-diffusion-model-from-stated-data%3A-Islam/db1e15f9adc914135f59e0d01ffc3cd051ac77f5.

Kabir, E., Kim K. & Szulejko, J. E. (2017). Social Impacts of Solar Home Systems in Rural Areas: A Case Study in Bangladesh. *Energies*, *10*(10)1615: 1–12. Retrieved from https://doi.org/10.3390/en10101615.

Kabir, E., Kumar, P., Kumar, S., Adelodun A. A. & Kim, K. (2018). Solar Energy: Potential and Future Prospects. *Renewable and Sustainable Energy Reviews*, *82*: 894–900. Retrieved from https://doi.org/10.1016/j.rser.2017.09.094.

Kojima, M. (2011). *The Role of Liquefied Petroleum Gas in Reducing Energy Poverty*. Extractive Industries for Development Series #25. Oil, Gas, and Mining Unit, The World Bank, Washington, DC. Retrieved from http://siteresources.worldbank.org/INTOGMC/Resources/LPGReportWeb-Masami.pdf.

Leaman, C. (2015). The Benefits of Solar Energy. *Renewable Energy Focus*, *16*(5–6): 113–115. Retrieved from https://doi.org/10.1016/j.ref.2015.10.002.

Lockton, D., Bowden, F., Greene, C., Brass, C. & Gheerawo, R. (2013). People and Energy: A Design-led Approach to Understanding Everyday Energy Use Behaviour. *Proceedings of EPIC 2013: Ethnographic Praxis in Industry Conference*. Washington, DC: American Anthropological Association. Retrieved from https://doi.org/10.1111/j.1559-8918.2013.00029.x.

Lofthouse, J., Simmons, R. T. & Yonk, R. M. (2015). *Reliability of Renewable Energy: Solar*. Institute of Political Economy. Utah State University. Retrieved from www.usu.edu/ipe/wp-content/uploads/2015/11/Reliability-Solar-Full-Report.pdf.

Marzolf, N. C., Cañeque, F. C. & Loy, D. (2015). *A Unique Approach for Sustainable Energy in Trinidad and Tobago*. Washington, DC: Inter-American Development

Bank. Retrieved from www.energy.gov.tt/wp-content/uploads/2016/08/A-Unique-Approach-for-Sustainable-Energy-in-Trinidad-and-Tobago.pdf.

Mathews, G. E. & Mathews, E. H. (2016). *Household Photovoltaics – A Worthwhile Investment?* Paper presented at the 2016 International Conference on the Domestic Use of Energy, Cape Town, South Africa. Retrieved from http://ieeexplore.ieee.org/stamp/stamp.jsp?tp=&arnumber=7466716.

Mazur-Stommen, S. (2013). *Ethnographies of Energy*. Retrieved from www.academia.edu/3489612/Ethnographies_of_Energy.

McGuire, G., Pantin, D., James D. & Seeterram, N. (2009). A Guide for Monitoring the Management of Oil and Gas Resources in Trinidad and Tobago. In *A Report of the Trinidad and Tobago Sustainable Development Network* (pp. 1–87). Retrieved from www.scribd.com/document/193973357/A-Guide-for-Monitoring-the-Management-of-Oil-and-Gas-Resources-in-Trinidad-and-Tobago.

Morris, P., Vine, D. & Buys, L. (2016). Residential Consumer Perspectives of Effective Peak Electricity Demand Reduction Interventions as an Approach for Low Carbon Communities. *AIMS Energy*, 4(3): 536–556. Retrieved from http://dx.doi.org/10.3934/energy.2016.3.536.

Muhammad-Sukki, F., Abu-Bakar, S. H., Munir, A. B., Yasin, S. H. M., Ramírez-Iniguez, R., McMeekin, S. G., Stewart, B. G. & Rahim, R. A. (2014). Progress of Feed-in Tariff in Malaysia: A Year After. *Energy Policy*, 67: 618–625. Retrieved from https://doi.org/10.1016/j.enpol.2013.12.044.

Narinesingh, D., Kumarsingh, K., Attzs, M., Gouveia, G., Beckles, D. M., Clarke, R., Chadee, D., Pantin, D. & Pollinais, S. (2013). *Second National Communication of the Republic of Trinidad and Tobago Under the United Nations Framework Conventional on Climate Change*. Government of the Republic of Trinidad and Tobago, Trinidad and Tobago. Retrieved from http://unfccc.int/resource/docs/natc/ttonc2.pdf.

Naspolini, H. F., Militão, H. S. G. & Rüther, R. (2010). The Role and Benefits of Solar Water Heating in the Energy Demands of Low-income Dwellings in Brazil. *Energy Conversion and Management*, 51(12): 2835–2845. Retrieved from https://doi.org/10.1016/j.enconman.2010.06.021.

NEED (National Educational Energy Development) Project. (2016). *Exploring Photovoltaics Student Guide*. Retrieved from www.need.org/files/curriculum/guides/Photovoltaics%20Student%20Guide.pdf.

Niu, S., Hong, Z., Qiang, W., Shi, Y., Liang, M. & Li, Z. (2017). Assessing the Potential and Benefits of Domestic Solar Water Heating System Based on Field Survey. *Environmental Progress and Sustainable Energy*, 37(5): 1781–1791. Retrieved from https://doi.org/10.1002/ep.12827.

Nolden, C. (2012). The Governance of Innovation Diffusion: A Socio-technical Analysis of Energy Policy. *EPJ Web of Conferences*, 33 (Article 01012): 1–8. Retrieved from https://doi.org/10.1051/epjconf/20123301012.

Pink, S. (2011). Ethnography of the Invisible: Energy in the Multisensory Home. *Ethnologia Europaea: Journal of European Ethnology*, 41 (1): 117–128. Retrieved from https://dspace.lboro.ac.uk/dspace-jspui/bitstream/2134/9407/2/S%20Pink_Ethnography%20of%20the%20invisible_2011.pdf.

Rallapalli, S. R. & Ghosh, S. (2012). Forecasting Monthly Peak Demand of Electricity in India: A Critique. *Energy Policy*, 45: 516–520. Retrieved from https://doi.org/10.1016/j.enpol.2012.02.064.

REC (Renewable Energy Committee). (2011). *Framework for Development of a Renewable Energy Policy for Trinidad and Tobago: A Report of the Renewable Energy Committee*. Trinidad and Tobago: Ministry of Energy and Energy Affairs. Retrieved from www.energy.gov.tt/wp-content/uploads/2014/01/Framework-for-the-development-of-a-renewable-energy-policy-for-TT-January-2011.pdf.

Rogers, E. M. (1983). The Innovation-Decision Process. In *Diffusion of Innovations* (3rd ed.) (pp. 163–209). New York: The Free Press.

Rüdiger, M. (2008). Introduction. In M. Rüdiger (Ed.), *The Culture of Energy* (pp. i–iv). Newcastle: Cambridge Scholars Publishing.

Ruotsalainen, J., Karjalainen, J., Child, M. & Heinonen, S. (2017). Culture, Values, Lifestyles, and Power in Energy Futures: A Critical Peer-to-peer Vision for Renewable Energy. *Energy Research & Social Science*, *34*: 231–239. Retrieved from https://doi.org/10.1016/j.erss.2017.08.001.

Sepp, S. (2014). Liquefied Petroleum Gas. In H. Volkmer (Ed.), *Multiple-Household Fuel Use: A Balanced Choice Between Firewood, Charcoal and LPG* (pp. 33–41). Deutsche Gesellschaft für Internationale Zusammenarbeit (GIZ) GmbH on behalf of the Federal Ministry for Economic Cooperation and Development (BMZ), Eschborn, Germany. Retrieved from www.cleancookingalliance.org/resources/287.html.

Sheller, M. (2014). Global Energy Cultures of Speed and Lightness: Materials, Mobilities and Transnational Power. *Theory, Culture and Society*, *31*(5): 127–154. Retrieved from https://doi.org/10.1177%2F0263276414537909.

Sorrell, S. & Dimitropoulos, J. (2008). The Rebound Effect: Microeconomic Definitions, Limitations and Extensions. *Ecological Economics*, *65*(3): 636–649. Retrieved from https://doi.org/10.1016/j.ecolecon.2007.08.013.

Southerton, D. (2012). Habits, Routines and Temporalities of Consumption: From Individual Behaviours to the Reproduction of Everyday Practices. *Time and Society*, *22*(3): 335–355. Retrieved from https://doi.org/10.1177%2F0961463X12464228.

Sovacool, B. K. (2009). The Cultural Barriers to Renewable Energy and Energy Efficiency in the United States. *Technology in Society*, *31*(4): 365–373. Retrieved from https://doi.org/10.1016/j.techsoc.2009.10.009.

Steg, L., Perlaviciute, G. & van der Werff, E. (2015). Understanding the Human Dimensions of a Sustainable Energy Transition. *Frontiers in Psychology*, *8*(Article 805): 1–17. Retrieved from https://doi.org/10.3389/fpsyg.2015.00805.

Strengers, Y. (2010). *Conceptualising Everyday Practices: Composition, Reproduction and Change*. Working Paper No. 6. Carbon Neutral Communities. Centre for Design, RMIT University, Melbourne, Australia. Retrieved from www.academia.edu/2076806/Conceptualising_everyday_practices_composition_reproduction_and_change.

Torriti, J. (2017). Understanding the Timing of Energy Demand Through Time Use Data: Time of the Day Dependence of Social Practices. *Energy Research & Social Science*, *25*: 37–47. Retrieved from https://doi.org/10.1016/j.erss.2016.12.004.

U.S. (United States) Department of Energy. (2003). *Just the Basics: Liquefied Petroleum Gas*. Retrieved from www1.eere.energy.gov/vehiclesandfuels/pdfs/basics/jtb_lpg.pdf.

Van Leeuwen, R. P., de Wit, J. B. & Smit, G. J. M. (2017). Review of Urban Energy Transition in the Netherlands and the Role of Smart Energy Management. *Energy Conversion and Management*, *150*(15): 941–958. Retrieved from https://doi.org/10.1016/j.enconman.2017.05.081.

Van Leeuwen, R., Evans, A. & Hyseni, B. (2017). *Increasing the Use of Liquefied Petroleum Gas in Cooking in Developing Countries.* Live Wire, World Bank Group. Retrieved from https://openknowledge.worldbank.org/bitstream/handle/10986/26569/114846-BRI-PUBLIC-add-series-VC-LWLJfinOKR.pdf?sequence=5&isAllowed=y.

Wallenborn, G. & Wilhite, H. (2014). Rethinking Embodied Knowledge and Household Consumption. *Energy Research and Social Science*, 1: 56–64. Retrieved from https://doi.org/10.1016/j.erss.2014.03.009.

Wallenborn, G. (2008). The New Culture of Energy: How to Empower Energy Users? In M. Rüdiger (Ed.), *The Culture of Energy* (pp. 236–254). Newcastle: Cambridge Scholars Publishing.

Wartluft, J. L. (1984). *Comparing Charcoal and Wood-Burning Cookstoves in the Caribbean.* Virginia: Volunteers in Technical Assistance. Retrieved from http://pdf.usaid.gov/pdf_docs/PNAAU242.pdf.

WEC (World Energy Council). (2016). *World Energy Resources- Solar 2016* (pp. 26–27). London. Retrieved from www.worldenergy.org/assets/images/imported/2016/10/World-Energy-Resources-Full-report-2016.10.03.pdf.

WEF (World Economic Forum). (2018). *Fostering Effective Energy Transition: A Fact-Based Framework to Support Decision-Making* (Insight Report). Geneva, Switzerland. Retrieved from www3.weforum.org/docs/WEF_Fostering_Effective_Energy_Transition_report_2018.pdf.

Winther, T. (2013). Space, Time, and Sociomaterial Relationships: Moral Aspects of the Arrival of Electricity in Rural Zanzibar. In S. Strauss, S. Rupp & T. Love (Eds.), *Cultures of Energy: Power, Practices, Technologies* (1st ed.) (pp. 164–176). Walnut Creek, CA: Left Coast Press. Retrieved from https://doi.org/10.4324/9781315430850.

Winther, T. & S. Bouly de Lesdain, S. (2013). Electricity, Uncertainty and the Good Life: A Comparison of French and Norwegian Households Responses to Policy Appeals for Sustainable Energy. *Energy and Environmental Research*, 3(1): 71–84. Retrieved from http://dx.doi.org/10.5539/eer.v3n1p71.

Yarbrough, I., Sun, Q., Reeves, D. C., Hackman, K., Bennett R. & Henshel, D. S. (2015). Visualizing Building Energy Demand for Building Peak Energy Analysis. *Energy and Buildings*, 91: 10–15. Retrieved from https://doi.org/10.1016/j.enbuild.2014.11.052.

Part III
Transitioning to and through household solar energy technologies

6 Electricity, solar hot water and mainstream cultural change

Introduction

The last chapter showed what the mainstream energy culture is like in a regime which has not mainstreamed solar energy technologies just yet. However, Barbados' residential electricity regime has mainstreamed SWHs, and based on the theory from Chapter 2, this means that it experienced an energy transition. But before delving into detail on solar hot water and mainstream culture change, it is important to give some characterization of the island's regime.

Like Trinidad, Barbados is an oil and natural gas producer. These resources were first discovered in the 18th century (Ince, 2018). However, Barbados only produces between 700 to 1,000 barrels of oil per day and between 290 to 500 BOE/day of natural gas which are shipped to Trinidad for refining and later imported (Gischler et al. 2009; Castalia Limited, 2010; SEforALL, 2012; Samuel, 2013; Ince, 2017, 2018).

Barbados imports over 9,000 BOE/day (Gischler et al., 2009; Castalia Limited, 2010; SEforALL, 2012; GoB, 2015; Thompson, 2015, 6) so this, together with the above figures, means that Barbados' fossil fuel production is significantly smaller than Trinidad's and the island is a net energy importer. The BNOC is the State-owned institution responsible for producing oil and gas as well as importing fossil fuels (Ince, 2018). However, the gas that is produced is put under the jurisdiction of the NPC (but both entities are being merged) (Ince, 2018).

Fossil fuels dominate the regime's energy mix and heavy fuel oil makes up 37%; diesel supplies 18%; gasoline accounts for 17%; other resources in the form of electricity give 14%; kerosene contributes 7%; bagasse is 3%; LPG provides 2% together with another 2% from natural gas; and solar water heating makes up less than 1% (Ince, 2018, 40). Interestingly and unlike Trinidad, natural gas is neither a large part of the energy mix nor the electricity mix. The regime's electricity mix is 74% heavy fuel oil; 17.4% kerosene; 5.6% bagasse; 2.2% diesel; 0.6% solar energy; and 0.2% natural gas (Ince, 2018, 41).

Barbados, like Trinidad, is serviced by one electricity utility company, the BL&P. The BL&P was established in 1899 and like the T&TEC, is a vertically integrated monopoly (Ince, 2018). However, unlike the T&TEC, the BL&P is privately-owned (Ince, 2018) and is responsible for power generation in addition to transmission and distribution.

The BL&P operates four power generation plants as well as purchases power from IPPs (BL&P, 2015a). Three of the four plants are large-scale centralized fossil fuel systems and the other is a 10-MW solar PV plant. Additionally, the IPPs' capacity largely refer to 14 MW of distributed PV systems. The regime's PV capacity will be discussed at the end of the chapter, but distributed PV is not part of the regime just yet. Nevertheless, the three fossil fuel plants total 252 MW which is a combination of 166 MW at the BL&P's Spring Garden facilities; 73 MW at Seawell; and 13 MW at the Garrison (BL&P, 2015a).

The FTC is the independent public utilities regulator on the island and acts in the same capacity that the RIC does in Trinidad. Other institutions involved in the wider electricity regulation and governance are governmental ministries such as the Ministries of Energy and Water Resources, Environment and Natural Beautification, and Finance, Economic Affairs and Investment; SWH retailers, for example Solar Dynamics and Solaris; as well as non-governmental institutions such as the BREA and UWI.

Relative to the residential sector, the BL&P has over 119,000 residential customers (Thompson, 2015; 2). The residential sector has made up the most of the electricity consumption and sales in recent history by accounting for 33% of the consumption compared to the 21% from commercial; 16% from public; 15% from tourism/hotel; 9% from industrial; and 6% from other sectors (Ince, 2018, 42).

At the household level, the 'average' Barbadian household consumes 358 kWh of electricity per month based on calculations using figures from Ince (2018, 80). This means that such a household spends an estimated US$86 on their monthly electricity based on the domestic tariff calculator provided by the BL&P (2015b) and using current rates.

As done in Trinidad, the costs of the fuels used to generate power are passed onto consumers. However, these costs and consequently the retail electricity rates fluctuate with the international oil prices because Barbados imports their power generation fossil fuels. The cost recovery is enabled through the FCA which allows the BL&P to pass on these costs. It is calculated monthly based on the sum of the previous month's renewable and non-renewable energy purchased, fuel consumed and any over or under recovery divided by the BL&P's kWh sales from the previous month (B&LP,[3] 2015). For example, in October 2018, the FCA rate was US$16 cents per kWh versus 14 cents in August 2019 (BL&P, 2015c).

The purpose of presenting this brief characterization of these aspects of Barbados' residential electricity regime is to show that whilst there are some fundamental differences between Trinidad and Barbados' regimes, for example use of primarily heavy fuel oil versus natural gas for power generation respectively, the regimes' structures share semblances, for example one electricity utility company and a reliance on large-scale, centralized fossil fuel power generation infrastructure. These sorts of semblances give the book's energy transitions narrative a sense of socio-technical continuity from one case study to another. So, it is with that that the adoption of SWHs can be looked it in more detail to fully complete the regime's characterization.

Adopting solar water heating

Knowledge and information

A good starting point to begin characterizing the adoption process is to look at what information residents wanted before adopting a new technology like SWHs (at the time). This relates to the first stage of the innovation-decision process, that is, knowledge (see Rogers, 1983 and Chapter 2). Table 6.1

Table 6.1 Showing the retail market's coding structure related to the types of solar hot water information that the residents interviewed in Barbados wanted before adopting their SWHs

Coding structure			Sub-code frequency
Theme/concept	Category	Sub-code	
Retail market	Systems in stock	Operability	1
		Effectiveness	1
		Reliability	1
		Retailer system durability	1
		Lifetime	1
		Capacities available	5
		Retailer system capacities available	2
		Retailer system types available	1
	Retailer services	Retailer system maintenance regimen	2
		Cost	1
		Retailer system capital cost	1
		Retailer system cost	1
		Retail cost package	1
		Market retailers	3
		Popular market retailers	1
		Cost vs. conventional electric water heating	1
	Other	Consumer experiences with retailers	1
		Household size impact	1
		Infrastructural house size impact	1
No information			5

Source: The author.

Note
The frequency is defined as the number of residents' answers that were tagged with the associated sub-code.

shows the types of information that the residents who were interviewed wanted at the time of adopting their SWHs. The table shows this as the sub-code labels which arose during the analysis and how they relate to each other in forming more substantive classes of information.

One observation to note is that there are five residents whose responses indicated that they did not search for any information prior to purchasing their SWHs, for example:

> None. We were building the house at the time and I went to a sale and it was offered for 500 dollars [Barbadian dollars]. You wouldn't take it at that price? I asked someone the average price for such a system just to compare prices and was told it is usually around 5,000 dollars.
>
> (Resident 26)

This is likely because residents forgot what information they sought out, and/or SWHs have become such a passive technology that it seemed like a natural thing to have adopted one. Therefore, SWHs may have been perceived as a common consumer good bought to meet lifestyle demands (Niu et al., 2017) – which is a sign of mainstreaming (see Chapter 2).

Nevertheless, the most salient information desired relates to the SWH system sizes that were available on the market at the time of adoption (see Table 6.1); the residents seemed more interested in the retailers' ability to provide adequately sized SWHs, for example:

> The biggest one [item of information] was the size of [storage] tanks available – how much hot water would it provide since it needs to be adequate enough for our usage. You don't want to have a case where the hot water runs out.
>
> (Resident 28)

As can be seen in the table, this is also set in the wider context of obtaining technical information such as the systems' operability; performance, that is, effectiveness and reliability; longevity, that is, durability and lifetime; types of systems available for sale; the costs involved; and finding out which were the retailers on the market.

Persuasion and motivations

The second stage of the innovation–decision process is becoming persuaded to adopt the innovation (see Rogers, 1983 and Chapter 2). It is useful to have an idea of what motivated the residents to adopt their SWHs because the power of choice is a key element of adoption and culture as shown in Chapter 3; human beings make decisions that are biased and reflect emotions and instincts (Gordon, 2011). The analysis pointed towards there being nine different potential motivators that persuaded the residents interviewed to adopt their SWHs.

The first is related to the personal dislike for cold water and shows the value of 'comfort', for example:

> My two young children complaining about cold water.
> (Resident 21)

The second motivator is based on the beliefs that using electric water heaters and pilot light heating systems to heat water have safety risks, as well as that using SWHs is safer. For example, with respect to electric water heating residents stated:

> after talking to people, many of them reported to have gotten shocked from electric water heating, so an electric water heater was not suitable.
> (Resident 21)

The resident who owned the pilot light heating system believed that:

> it is also not safe having an open fire in the house.
> (Resident 17FBDS)

These views positioned 'safety' as another motivator.

Third, residents value 'durability'. Several expressed displeasure with the fact that the electric water heaters' elements, the component that heats the water, are fragile and in need of regular replacements. For instance, this was a nuisance because:

> You always have to buy replacement elements for electric water heaters.
> (Resident 18)

One other resident also said that their parents had to replace their electric water heater's element every six to seven months. Further, another compared their SWH to the electric water heater's fragile elements and pointed out that by comparison, their SWH has a regular service regimen (every five years) which suggests that it was perceived as more durable.

'Reliability' is the fourth motivator based on the beliefs that the sun is an infinite resource so is a guaranteed heat source, and more explicitly, alternatives such as the aforementioned pilot light water heating systems are not believed to be dependable or reliable.

A fifth motivator arose as 'convenience' through beliefs tied to electric water heating requiring some time before providing hot water; pilot light water heating and LPG stoves simply being perceived as being inconvenient; and the convenience of one-time capital cost payments for SWHs, for example:

> Boiling water on the stove was an inconvenience.
> (Resident 21)

130 *Transitioning through household solar*

> You pay a lump sum for your solar water heater but after doing that you don't have to worry about that [paying for the system] again.
>
> (Resident 31)

Related to 'convenience' is 'practicality' as a sixth motivator. It is drawn from the belief that the hot water provided by several of the residents' prior electric water heaters did not provide hot water to where it was needed, for example:

> Hot water is used for washing dishes and the electric water heater doesn't accommodate this.
>
> (Resident 24)

As stated in Resident 24's quote, their electric water heater was not capable of providing hot water where they needed it, that is, in the kitchen to wash dishes. This was also complemented by another resident's more direct belief that solar water heating is, quite simply, practical.

The seventh motivator is attached to the beliefs that using electric water heating increases the electricity bill and is expensive – though it is dependent on how often it is used. In addition, using SWHs was believed to be cheaper than electric water heating and conserves electricity. Such beliefs point towards residents valuing 'cost-effectiveness'. Examples include:

> electric water heating makes your electricity bill high.
>
> (Resident 31)

> Electric water heaters are expensive.
>
> (Resident 32)

Finally, residents value 'social validation' and 'awareness' as the eighth and ninth motivators. Residents reported valuing the word-of-mouth from peers, seeing SWHs that have been installed by peers and seeing television advertisements. The best example of a supporting quote is:

> It [solar water heating] was introduced on television for a period of time, people bought it, and I also heard about it from some of them.
>
> (Resident 25)

Figure 6.1 shows the motivators that persuaded the residents to adopt their SWHs: comfort, safety, durability, reliability, convenience, practicality, cost-effectiveness, social validation and awareness were motivators for technologically transitioning to solar water heating. However, the convenience and comfort provided by using solar hot water are the most salient.

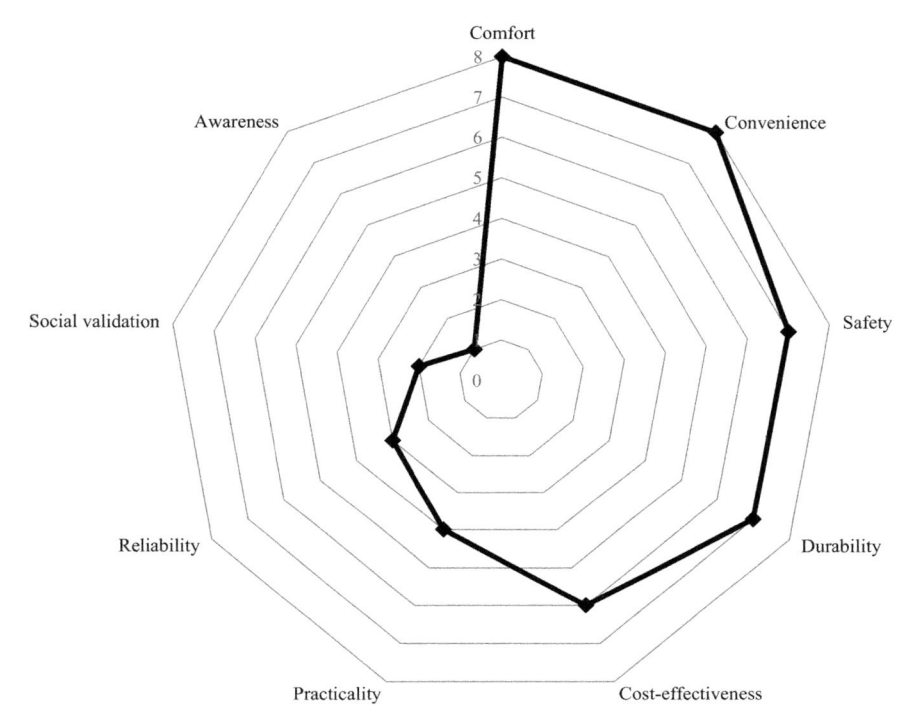

Figure 6.1 Showing the relative significance of the motivations for adopting solar water heating based on the inductive thematic content analysis of the interviews conducted with residents in Barbados.

Making the adoption decision and implementation

Many industrialized countries established solar water heating industries by the 1980s (McVeigh, 1984). But Barbados' had its origins in the 1960s, and it took off when Solar Dynamics was established in 1974/1975 as well as when its first major rival, SunPower, entered the market in 1978 followed by a third, Aqua Sol (which later became Solaris), in 1981 (Bugler, 2012; Rogers et al., 2012; Samuel, 2013; CDB, 2014; Husbands, 2016).

The Government also implemented several policies to promote the adoption of SWHs, for example in 1974, 20% tax breaks were offered for SWH manufacturing; 30% tax on electric water heaters; tax refunds for eligible SWHs for homeowners; and legally mandating solar hot water for new build governmental housing (Bugler, 2012). In addition to these policies and the establishment of an indigenous manufacturing and retail industry, retailer-led marketing and advertising (specifically from Solar Dynamics) and the 1970s oil price hikes were also factors that made solar water heating in Barbados an attractive economic option (Bugler, 2012).

The solar water heating market was dominated by the three companies well into the 2000s but Solar Dynamics had 55 to 60% of the market share and the remaining was relatively evenly distributed between SunPower and Aqua Sol/Solaris (Perlack & Hinds, 2003; CDB, 2014). Solar Dynamics was and is the market leader in solar water heating (ECLAC, 2003).

The oldest SWH owned by any of the residents interviewed was adopted in 1983 and the most recent was in 2014. In the wider industry, SWH adoption increased from 12 systems in 1974 when the industry 'started', to 1,848 in 1980; 19,370 in 1990; 31,000 in 1999; between 32,000 and 35,000 in 2002; and between 45,000 to over 50,000 in 2009 (Jensen, 2000; UN, 2002; Perlack & Hinds, 2003; Bugler, 2012, 1; Husbands, 2010; NREL, 2015a). Getting more recent figures proved challenging but thinking about the SWH penetration lends perspective.

Nearly 75% of the SWHs in Barbados are residential (Moore et al., 2014, 6). Sources such as the CEIS (2015), Gray et al. (2015), and the IEA-ETSAP and IRENA (2015, 1), however, report that 80 to 90% of households now have solar hot water. Given these figures, and when put in the context of the mainstreaming pseudo-threshold of 16% outlined in Chapter 2, it seems reasonable to suggest that the energy transition which resulted from mainstreaming SWHs would have caused an energy culture shift in Barbados' mainstream energy culture – which would have arguably been more similar to Trinidad's before.

So, having presented this regime-level overview of SWH adoption, it is worth highlighting how the residents who were interviewed got their SWHs since it would give an in-situ qualitative appreciation for the state of the regime's solar water heating industry. It will also capture the holistic innovation-decision process (see Rogers, 1983 and Chapter 2) to see what organic elements of adoption emerged; and it sets the wider context for the knowledge and persuasion stages looked at before by giving insight into the residents' views related to the decision, implementation and confirmation stages of their adoption of solar water heating. There are five different adoption pathways based on the residents' accounts.

The most popular and common option involved becoming familiar with the solar water heating retail market and its retailers; liaising with these businesses on investment options; settling on a retailer; signing up with the company; agreeing on a payment plan (down payments or full upfront payments); and finally, having the system installed. Examples of responses that add contextual summaries are:

> I looked to see who were the main players, then engaged people who have had experience with these players before buying it, and then spoke to the popular retailers relative to quotations; installation; maintenance; savings and discounts; as well as installation time.
>
> (Resident 17)

> I can't remember exactly but I know at some point, I would have went to the company; filled out their forms; selected an available payment plan; made the down-payment; then came the installation; and finally paying-off for the system.
>
> (Resident 19)

Interestingly, with respect to choosing a retailer, this was linked to the business' popularity and establishment, and the impression given by residents was that popularity and establishment are linked to retailers' numbers of systems sold, size of customer-base and length of time they have been in business, for example:

> I asked around from companies. [I] wrote down questions I had and sought the answers from them – and eventually chose the company that was around the longest.
>
> (Resident 21)

The retailers' cost-effectiveness was another factor. Whilst there are inherent values that make more popular, established and cost-effective retailers the best option for residents, in some cases, the retailer was only chosen because they were believed to be the only ones on the island at the time, for example:

> [I] Called up the retailer, made a down payment, [and] they came and installed it. At that time there was only one retailer – and besides, all the brands that we have today serve the same function.
>
> (Resident 25)

> When we got it, Solar Dynamics was the sole provider at the time then the others came in after.
>
> (Resident 27)

Based on the above, at the time of all the residents acquiring their different systems (between 1983 to 2014) there was more than one company operating. Therefore, the fact that residents thought that there was only one company can perhaps be due to retailers' marketing and extent of visible brand adoption at the time. Perlack and Hinds (2003) suggest that Solar Dynamics' marketing was a huge driver of the early SWH industry, and the company also has the largest portion of the market. Therefore, it is easy to see why residents could believe that there was only one company operating at the time if the market leader's promotions were so dominant.

A second adoption pathway is through door-to-door sales representatives. One resident outlined that a salesperson offered them a rent-to-own proposal and after three years of 'renting', owned the system. Interestingly this resident even acknowledged that:

> most people also usually just go to SunPower or Solar Dynamics.
>
> (Resident 24)

The third option is through non-manufacturer retailers (versus manufacturer retailers e.g. Solar Dynamics). One resident outlined that they got their SWH through a hire purchase plan with a non-manufacturer-retailer; another did so more informally by visiting a garage sale and bought a second-hand SWH; and one other compared quotes from a private individual and reputable company, weighed the pros and cons of each, then chose the private installer.

A fourth adoption narrative is linked to the real estate market because one resident reported buying their present home (which they have been residing in since 2013) with the SWH having been already installed. The fifth option is through social networks and relationships because a resident simply said that they got their SWH through a family member who works in the SWH retail business.

These five adoption pathways include several norms, for example there is only one company on the island; practices, for example completing installation permits; material culture, for example payments; and external influences, for example retailers' marketing and advertisements. These features of the incumbent mainstream energy culture already begin to differentiate the residents interviewed in Barbados from those in Trinidad. However, whilst the features are part of the mainstream energy culture, the impact of using a SWH should now be in the spotlight.

To do this, the previous mainstream energy culture needs to be compared with this present one to highlight the changes linked to SWHs – and this shows the value of looking at the mainstream energy culture from Trinidad in Chapter 5. So, the following sections are dedicated to pointing out the mainstream cultural changes associated with the adoption of SWHs and what a mainstream (solar) energy culture looks like after the mainstreaming of SWHs.

Solar hot water and mainstream cultural change

Energy supply

Figure 6.2 shows the households' energy inputs based on the residents' statements which constitute their 'energy supply' cultural domain. If this figure is compared to Figure 5.1, electricity and LPG are the common household energy inputs. However, batteries and, most notably, solar energy are also acknowledged by the residents interviewed in Barbados. Adopting and using solar water heating therefore had a cultural impact on the perception of where/how households get their energy.

The electricity and LPG references are quite like those given by the residents from Trinidad in Chapter 5 so in the interest of limiting redundancy as well as conciseness, they will not be elaborated on. With respect to the use of batteries, one resident simply stated that batteries powered some of their appliances and did not specify the type(s) of batteries being referred to. Therefore, the key element is the shift in the household energy mix of the

Electricity, solar hot water and mainstream cultural change 135

SOLAR ENERGY
Total Sub-code Frequency = 25
Total Number of Sub-codes = 12

ELECTRICITY
Total Sub-code Frequency = 23
Total Number of Sub-codes = 24

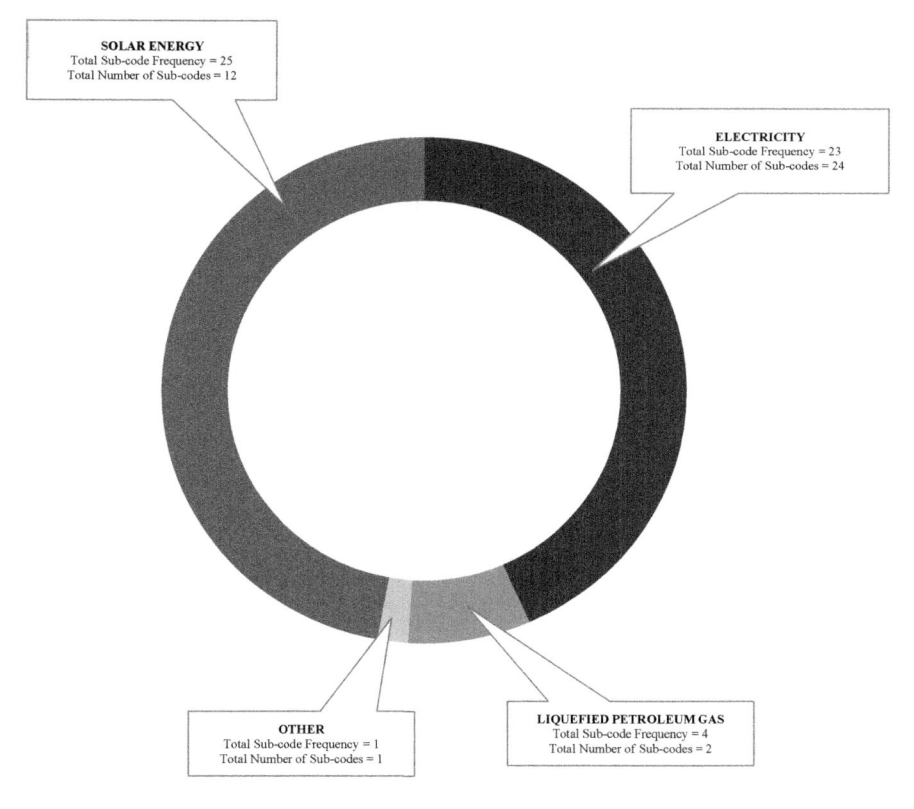

OTHER
Total Sub-code Frequency = 1
Total Number of Sub-codes = 1

LIQUEFIED PETROLEUM GAS
Total Sub-code Frequency = 4
Total Number of Sub-codes = 2

Figure 6.2 Showing the household energy inputs which emerged from the inductive thematic content analysis of the interviews conducted with residents in Barbados.

mainstream energy culture to include solar energy. Examples of residents' statements to highlight this key difference between Figures 6.2 and 5.1 are:

> Practically, most, about 90% comes from [the] Barbados Light and Power [company] and 10% from solar water heating for the kitchen and bathroom.
> (Resident 21)

> A SunPower water heater for bathing and washing dishes, electricity, and gas for the stove i.e. a bottle/tank of gas – I only have three sources.
> (Resident 31)

Table 6.2 shows the raw coding structure for 'solar energy' as a category in the analysis which was based on the residents' responses (such as the above quotes). The residents not only connect with the sun as the resource, but also

Table 6.2 Showing the solar energy category's coding structure which forms part of the 'energy supply' cultural domain

Coding structure				Sub-code frequency
Category	Sub-category	Code	Sub-code	
Solar energy	Solar energy is a domestic energy input			9
	Decentralized solar energy technology	PV	PV facilitates solar's utility as a domestic resource	1
			PV involve electrical energy	1
			Ambiguity in how photovoltaics work	1
		SWHs	SWHs facilitate solar's utility as a domestic resource	6
			SWH brand recognition	1
			SWH location reference	1
	Solar technology operates within defined energy dimensions	Defined energy space for solar technology	Solar hot water used in defined energy spaces	1
			PV used in defined energy spaces	1
		Defined energy service provision from solar technology	Solar hot water used for defined energy services	1
			PV used for defined energy services	1

Source: The author.

Note
The frequency is defined as the number of residents' answers that were tagged with the associated sub-code.

with the utility of PV and solar water heating – though PV was mentioned by only one resident who owned outdoor solar lighting in addition to their SWH.

Nevertheless, it is worth pointing out that based on the solar references given in residents' responses, the solar technologies mentioned operate within defined energy spaces and are used for defined practices. In the case of solar water heating, the technology provides hot water for bathing, or washing dishes and/or clothing (practices) that take place in the bathrooms or kitchens of homes (energy spaces); the previous quotes also provide contextual examples for this.

Energy usage and its costs

As noted in Chapter 5 for the residents interviewed in Trinidad, both their 'energy usage' and 'energy costs' domains overlap in terms of their material culture, norms, practices and external influences – and the same is true in the

Electricity, solar hot water and mainstream cultural change 137

case of the residents interviewed in Barbados. Further, the mainstream energy culture samples in both islands are quite similar. Examples of quotes which provide comparable references to those given by the residents in Trinidad are:

> As a retiree, we are home all day so the fridge [refrigerator], TV [television] and radio are on all day long.
>
> (Resident 18)

Further, Figures 6.3 and 5.2 show a similar pattern in the perceived peak energy demand time slots chosen by residents, that is, there is a common belief that the most energy is consumed by the household in the evening/night. This is also inclusive of similar material culture, norms and practices driving these time selections such as the use of lighting, refrigerators and televisions. Once again, material cultures were the most salient elements given the frequent references to household appliances.

Residents converged on the 7:00 pm to 8:00 pm slot as the specific perceived peak energy demand time within the broader 5:00 pm to 11:00 pm peak shown in Figure 6.3 (which is comparable to the 6:00 pm to 8:00 pm selections in Trinidad from Chapter 5). In the wider context, Barbados' grid technically experiences its peak demand roughly between 9:00 am to 4:00 pm and again from 6:00 pm to 8:00 pm (Hohmeyer, 2015; Rogers, 2016); this is likely due to the consumption in places such as schools and offices during the day, and when persons return to their homes in the evening.

However, despite the strong similarities based on the residents' interviews in Trinidad and Barbados, there are three major distinctions between both samples that are worth pointing out.

The first is the external influence of climate change and the availability of water on energy use. One resident flagged climate change as a key external influence on their energy usage because in their opinion it affected the use of cooling material culture, for example:

> with climate change and the heat, air conditioning use is key.
>
> (Resident 25)

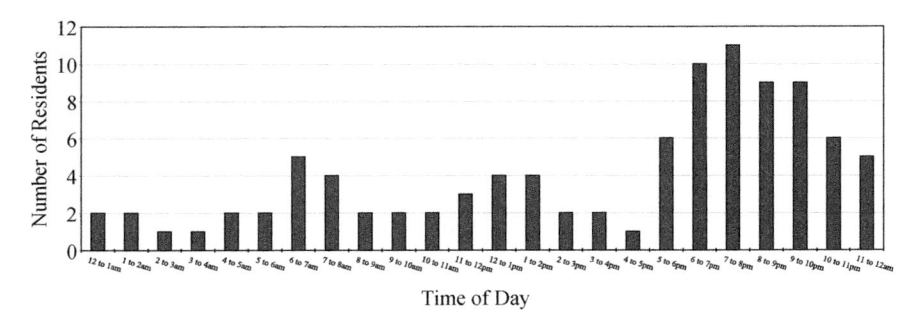

Figure 6.3 Showing the perceived peak energy demand times of the residents interviewed in Barbados.

With respect to water availability, another resident outlined that they recently installed a tank and water pump because they live in a water-scarce region of Barbados. This is worth highlighting because it can potentially be linked to the water input needed for using their SWH.

Second, there are some key interactions between material culture, norms and practices noted in the responses given by residents in Barbados which are associated with energy conscientiousness and the costs of energy. The references given relate to energy-efficient appliances as well as practicing energy conservation. With respect to the latter, the practices involved unplugging appliances when not in use, switching off appliances when not needed and using energy-saving appliances, for example turning off fans when not in use, installing timed fluorescent lighting and comparing electricity bills before and after getting into the habit of unplugging appliances that are not in use, respectively. For instance:

> I tried a test by unplugging everything when I leave [my] home and the bill has dropped e.g. unplugging the cell phone['s] charger, fan, microwave, [and] electric kettle.
>
> (Resident 23)

> Turning off the fans however, that decreases it [the electricity bill]. Leaving appliances plugged-in also increases the bill.
>
> (Resident 26)

Third is the external influence of cloud cover and the influence of the international oil markets on the FCA. Relative to cloud cover, one resident stated that their solar hot water supply is affected by cloudy skies during December and January when they believe cloud cover to be at its highest; cloud cover reduces the amount of incident solar energy on the SWH's collector surface and consequently reduces the water heater's efficacy (Curry, 2007; Goss, 2010). This observation is important because this was the only time that any reference to solar energy emerged organically in the energy usage or cost domains (and will be elaborated on later).

In the case of the FCA, two other residents pointed out its implicit influence: one resident simply highlighted the influence of the international oil market and specifically noted the fluctuant oil price's impact on their electricity costs, and the other believed that there are potential loopholes in the FCA cost recovery mechanism which enables the utility company to have a significant amount of institutional control over electricity prices.

Nevertheless, though the spread of the residents' perceived electricity bills in both Trinidad and Barbados are not that different (US$25 to 175 in Barbados versus $43 to 180 in Trinidad), in Barbados, the influence of the FCA and international oil markets is likely responsible for the more varied opinions on residents' electricity bills being either too high or reasonable/affordable. This is likely because the fluctuating prices mean there is no consistent cognitive baseline that could be formed surrounding what an 'average' bill is since using the same amount of energy in a given billing cycle would result in a different bill in another because of the FCA. For example, residents said:

> depends on the price of oil since the market price fluctuates.
> (Resident 25)

> Because the bill went up due to an increase in the fuel rate.
> (Resident 28)

Recalling that one resident flagged cloud cover as an external influence on their solar hot water supply, the interviews included a more thorough and explicit series of questions to gauge the impact of SWHs. There was one response which showed that the resident was completely unsure about the technology's impact. But the other residents were split between there being no perceived impact on their energy usage and its costs, and there being an impact on both.

Interestingly, only two residents of the eight who thought there was no perceived impact outright believed so. The other six all had norms and practices which pointed towards there being an energy change that was attributable to using solar hot water, for example:

> No real impact on the home's energy use really. Although I would add that when there's no sun and you switch to electrical [via the SWH's backup booster] to heat water, because it ties into the electrical panel, it would increase the bill and ultimately consumption as well.
> (Resident 17)

> It has no impact on the bill. [I] Never really used the booster [referring to the SWH's electrical backup] since in Barbados … it is always sunny and getting two cloudy days is rare.
> (Resident 25)

There are three possible reasons as to why there may have initially been no perceived impact by solar water heating.

The first is that in Barbados, the global oil price fluctuations are reflected in the retail electricity rates (NREL, 2015b; Hohmeyer, 2015). So, there may not be a consistent cognitive baseline for what SWHs' impact on electricity bills at the household level is; the bill fluctuations are subject to landscape-level factors beyond the home's energy consumption.

Second, since SWHs are generation technologies this means that they could have been perceived as being outside the conventional consumptive framing of household energy. As will be shown later, SWHs are touted as having reduced several residents' household energy consumption. Therefore, the systems are believed to conserve electricity. However, recalling that material culture like household appliances are the most salient type of material culture in the mainstream energy culture from Chapter 5 as well as this one, household appliances are consumptive. But SWHs are generators so this can be a reason why some did not perceive any impact on their energy usage or costs.

The third and final reason is that as the residents became more familiar with the technology, it became less 'observable' (Labay & Kinnear, 1981). Cressman (2009) states that because a technical element is functioning as it should, the complexities of it get lost – similarly to electricity infrastructure which was once socially contested but is now invisible because of its passive infrastructure (Richerson, Mulder & Vila, 2001; Namias, 2008; Sovacool, 2009; Schelly, 2014; Wallenborn & Wilhite, 2014).

From the earlier quotes, the impact-stimulus alluded to is the use of the system's electrical auxiliary heater which enables the water to be heated using electrical energy instead of solar by flipping the associated switch (the practice). The residents who outright connected with an impact thought that their SWHs affected their household's use of and costs of using LPG and electricity (the norms).

As detailed in the last chapter, LPG is used as a cooking fuel (McIntyre et al., 2016) and is usually connected to a stove like the one shown in Figure 6.4. There is also usually a steel cylinder, a pressure regulator, a hose that connects the regulator to a burner, and a burner (Sepp, 2014). Relative to the impact on LPG, two examples of responses given are:

> No real impact. If I didn't have it [referring to the SWH], I'd be using gas [to heat water on the stove] to wash the dishes.
>
> (Resident 18)

Figure 6.4 Showing water being heated on a stove by combusting LPG.

Well for sure ... if I was using an electric [water] heater the electricity bill would go up and the same would go for using a gas stove in terms of the costs of buying more gas.

(Resident 28)

The residents therefore referred to the use of LPG to heat water on their stoves where its costs and consumption would have been higher had it not been for solar water heating. Combusting LPG for heat energy was substituted for solar energy, and work such as Headley (1998), Perlack and Hinds (2003), Milton and Kaufman (2005) and Marzolf, Cañeque and Loy (2015) touch on the LPG to solar substitution.

In terms of the impact of solar hot water on electricity, examples of responses are:

It keeps it [referring to the electricity bill] lower because the switch [for the electrical backup booster] isn't used at all since it doesn't get cold enough to require using the switch to enable electrical heating of the water when solar energy is limited.

(Resident 24)

During rainy months, when the booster [referring to the solar water heater's electrical backup] is turned on, it increases the bill – especially if you forget it on.

(Resident 27)

What these examples suggest is that solar hot water is perceived to have either reduced energy bills and consumption because of its adoption and subsequent usage, that is, electrical energy is substituted for solar energy, or the home's energy consumption and bills increased because the water heater's electrical auxiliary heater was used, that is, solar energy is substituted for electricity.

As shown by the extracts from the residents' accounts, using domestic solar water heating displaces the need for other heating fuels and electricity (NREL, 1996; Milton & Kaufman, 2005; Del Chiaro & Telleen-Lawton, 2007; READ, 2014). It saves households money and energy by reducing the consumption of energy supplied by the incumbent system(s) and so too the costs of consumption (Headley, 1998; Milton & Kaufman, 2005; Hill et al., 2011; Kakaza & Folly, 2015).

This section illustrates that SWHs have been integrated into the mainstream energy culture as material culture; acquiring the SWH and flipping its electrical backup switch are practices; and associated norms are the perceived energy costs and consumption reductions that follow from adopting and using solar hot water, and vice versa, when the electrical auxiliary heater is used. The emergence of new habits is thought to indicate energy culture shifts (Stephenson, 2012). So, after comparing the samples of the mainstream energy cultures looked at here and in Chapter 5, the culture of Barbados underwent a cultural shift.

142 *Transitioning through household solar*

The cultural cognizance of solar energy substituting for LPG and electricity also points towards a cultural transition because the residents' household energy input norms have changed to include solar energy (as shown in the preceding section). These thoughts consequently show a cultural consonance between the impact of SWHs on the energy supply, on the cultural constructs of energy supply, usage and costs since they affect 'electricity' and 'LPG' in all three.

But interestingly, since the energy usage and costs domains are dominated by electric-oriented features, the practice of flipping the electrical auxiliary switch means that the SWH is positioned outside this main electrical cultural construct – much like the way in which household solar energy technologies have tended to play a disruptive role outside the conventional, technical energy system (Manning, 2015). Therefore, electricity acts as a connector across the cultural transition space.

Solar energy and solar water heating

There is even a comparable cognizance of solar energy as a resource in both Trinidad and Barbados' mainstream energy culture samples because the norms tied to solar energy are centred on it being a resource, requiring a technology to be used and having some relative advantages over the incumbent and possible alternatives, for example:

> Well I would say that it's some form of alternative energy not fossil fuel.
> (Resident 19)

> Renewable energy that we get from the sun.
> (Resident 27)

Relative to the technology (the SWH), there was also a comparable core understanding based on the norms the residents associate with the technology. Therefore, both islands' residents' norms have some conceptual consistency between the notions of 'solar energy' and its application through the SWH, that is, a resource and technology consistency because the role of the sun is the key feature of both. Further commonalities are the fact that SWHs are affected by the diurnal availability of sunlight; use a solar collector surface to capture sunlight; enable heat energy transfers; circulate water; and store water in storage tanks. For example, residents in Barbados said:

> The sun's heat is converted to heat to heat the water in the storage tank.
> (Resident 28)

> It is installed on the house. The hot water pipes are connected to the cold pipes which take it to the tank and the sun generates a certain amount of heat during the day to heat the water.
> (Resident 31)

Nevertheless, there is one key distinction that separates Barbados' residents' perceptions from those in Trinidad: the SWH can supplement its heating through conventional electric water heating during periods of limited sunlight by flipping the associated electrical auxiliary switch brought up in the last section. This shows that the residents' norms, for example the belief that cloud cover (an external influence) reduces the SWH performance, are manifested as a practice, that is, flipping the switch. For example, residents said:

> But if rainfalls ... you use the electrical booster by flipping the switch.
> (Resident 22)

> If you have 2 cool days without sunlight you can use the electrical booster to heat the water.
> (Resident 23)

This observation is quite significant because it supports the findings from the preceding section which showed that the mainstream energy culture in Barbados experienced a cultural shift related to the norms and practices associated with (not) flipping the electrical backup switch that were developed post-SWH adoption.

Transitioning from solar water heating to photovoltaics

The chapter's introduction did not include any substantial information on Barbados' household PV because it is still niche level. So, in that light, the chapter's writing here continues the book's underlying energy transitions narrative where Barbados' regime which was originally wholly structured around centralized fossil fuel power generation infrastructure transitioned to include SWHs and is now in the process of transitioning to/through PV.

Even though SWHs are the case study technologies being looked at in Barbados, referring to the island's PV is an apt way to end the chapter because it is important to highlight the solar energy transitions unfolding in the island; to bring a sense of continuity and completion to the Barbadian solar story included in the scope of the book; and to acknowledge that PV will be building upon the success of SWHs on the island (Buchinger et al., 2018; Ince, 2018).

All the PV installed in Barbados prior to 2010 were standalone systems except for two distributed systems which were installed by the BL&P in 2001 (Rogers, Chmutina & Moseley, 2012). Historical examples of standalone installations in the early 2000s are: 60 kWp at Harrison's Caves; 50 kWp at the BNOC's main offices; 50 kWp on Government buildings; 26 kWp on roughly 20 homes; 11.1 kW at the Skeete Bay fishing complex; 3 kW at Combermere School; 1.1 kW at the UWI; and a 300-W mobile public demonstration system (Rogers, Chmutina & Moseley, 2012; 5; Buchinger et al., 2018; 25). Therefore, distributed PV has only been around in Barbados for roughly nine years.

During these early 2000s, there was an estimated 200 kWp of PV installed on the island of which 55% was commercial, 31% was public and 15% was residential (Rogers, Chmutina & Moseley, 2012; 5) – which can be compared to the island's renewable power generation ambition that hopes to achieve a 75% reduction in fossil fuel usage which should be replaced by renewables and of which 15%, that is, almost 195 MW, should come from solar energy (Buchinger et al., 2018, 26; Ince, 2018).

But 2010 was a significant year for PV in Barbados because one of the most significant solar-related policy mechanisms was instituted to support PV, that is, the RER. The RER was a large part of Barbados' distributed PV implementation (NREL, 2015b; Ince, 2018) to the point where there were less than ten systems installed around the time it was implemented and within five years this grew to over 300 (Buchinger et al., 2018, 24).

The RER was instituted as a two-year pilot scheme (2010 to 2012) for connecting PV to the grid (Rogers, Chmutina & Moseley, 2012; NREL, 2015b). Under the RER, the BL&P purchases power from the owner of an enrolled PV system at a rate determined by the FTC whilst the owner of the PV system is simultaneously charged at current rates for the electricity consumed from the grid over a ten-year contact (BL&P, 2014).

The RER was set at 1.5 times the customer's average consumption, up to a maximum of 150 kW per system, and the credit rate was set at 1.8 times that of the FCA (Rogers et al., 2012); the FCAC is a per kWh rate which reflects the fluctuations in the international market's oil price on the local electricity prices in Barbados. This means that when oil prices are high, the RER rate is high and vice versa (Buchinger et al., 2018).

Enrolling in the program was so done on a first come, first served basis with a customer cap of 200 or installed capacity cap of 1.6 MW depending on which maxed out first (Rogers, Chmutina & Moseley, 2012). Further, if the system being enrolled is greater than 2 kW in capacity then the customer will be billed under a 'Buy All/Sell All' arrangement where they are billed by the BL&P for all the electricity consumed (regardless of the source) and credited on the same bill for all the power generated from the system enrolled in the RER (BL&P, 2014).

But if the system is smaller than 2 kW then the customer can choose between either this or an alternative known as 'Sale of Excess' where they are billed by the BL&P for what is consumed from the grid and credited for the excess electricity sold (BL&P, 2014) – the latter aspect means that the customer only exports and sells the excess that they do not consume from their PV system.

After the end of the pilot, the RER was expanded in 2013 and it was reported that by the end of that year, 271 systems were enrolled (1.7 MW) and at the start of 2014, this rose to 424 (3.5 MW), which amounted to just about half of the new 7 MW cap for the program (NREL 2015b; 3). However, in early 2017 there was an estimated 14 MW of IPP distributed solar installed (BL&P, 2015a; BloombergNEF, 2018).

The original 150 kW individual system cap was also been increased to 500 kW (BL&P, 2015d) and the former credit rate of 1.8 times the FCA was changed to 1.6 (BL&P, 2014; Buchinger et al., 2018; BloombergNEF, 2018). The RER credit rates are reviewed periodically by the FTC (BL&P, 2014) and sources such as the FTC (2016) and Madden (2019) report that in light of recent reviews, there are calls to move towards a permanent credit rate under the RER in order to meet the island's overall renewable energy ambitions.

It is worth mentioning that Barbados did not have any utility-scale PV (NREL, 2015b), that is, installations connected at the transmission level of the electricity grid versus the distribution level which is where systems like household PV such as those under the RER would be connected. However, this changed in 2016 when the BL&P established a 10-MW system on the island (Buchinger et al., 2018, 24). The solar plant covers about 170,000 m^2 and has just under 44,500 PV panels (BL&P, 2015a). This plant is also likely the reason for the massive jump from the 3.5 MW of PV to 14 MW quoted earlier between 2014 and 2017.

Whilst the RER has been arguably the main policy driver of distributed PV to date in Barbados, other supporting initiatives include RETs of 100% renewable power generation by 2045 and 15% of Barbados' energy to be generated by solar by 2037 (Buchinger et al., 2018, 26; Ince, 2018); the waiving of the import duty and environmental levies for PV systems; and as are applicable to SWHs, tax incentives which allow homeowners to reclaim their income tax on the money invested in an eligible PV system (Rogers, Chmutina & Moseley, 2012).

Whilst this brief section is not exclusive to residential PV, it still shows that in Barbados' wider energy system, there are developments which can mark the initial phases of a transition to/through PV. However, the next chapter illustrates such a transition through PV in the island of O'ahu.

Summary

Barbados' residential electricity regime is quite like Trinidad's. For instance, Barbados and Trinidad are serviced by one electricity utility company each and are dependent on large-scaled, centralized power generation infrastructures which run on fossil fuels. However, some key differences include the primary reliance on heavy fuel oil for power generation versus natural gas respectively, and most notably, the influence of household solar energy in the form of household solar water heating as well as PV in Barbados – though the latter is not yet part of the regime.

With respect to SWHs, at the time of adoption, they would have been perceived as innovations. Therefore, it can be expected that before adopting such an innovation, residents would want to find out more about it. Information related to technical aspects such as durability and operability as well as market services such as maintenance regimens are amongst the salient information desired before adoption. However, what was seemingly the most prominent based on the interviews with residents in Barbados was information on retailers' ability to provide appropriately sized SWHs. Further, the comfort and convenience afforded by using solar hot water are the strongest motivators for adopting an SWH based on the interviews conducted.

Once residents were satisfied with their knowledge about solar water heating as well as becoming persuaded to adopt it, next came the actual adoption. Based on the interviews, there can be combinations of either manufacturer or non-manufacturer SWH retailing as well as formal and informal retailing. However, the mainstream way is to select a preferred manufacturer-retailer and then liaise with them on the administration, payments and installation.

This combination of adoption pathways means that there are currently no well-defined niches for adopting solar water heating anymore – which is what should happen once a technology is mainstreamed. This also suggests that increasing the possible ways of adopting solar energy technologies should be encouraged. However, this should come with some quality control to ensure that standards are upheld so that the market is guarded against inferior products and services.

Barbados' underlying mainstream energy culture shares strong similarities with Trinidad's. This is because the energy cultures have been formed around the provision of electricity from a single utility company as well as the institutions and infrastructure built around that. So, both mainstream energy cultures share similar material culture, for example refrigerators and televisions; practices, for example cooking and showering; norms, for example the perceived peak energy demand occurring during evenings/nights; and external influences, for example the tropical heat and workplace schedules.

However, there are some notable differences such as the influence of the international oil prices and the FCA on electricity bills in Barbados. But the most observable differences between the mainstream energy cultures sampled in both islands are attributable to the adoption and use of solar water heating. The norms and practices associated with acquiring SWHs are defining differences because they relate to physically getting the technology for the home.

Adopting and using an innovation transforms it from being technology into material culture because people's interactions with it as part of their lifestyle gives it cultural meaning. Adopting and using technological innovations also increases users' understanding of it because the user develops 'lived-in' insights into how it works.

As a result of such 'lived-in' experiences, adopting solar energy technologies like SWHs will catalyse a cultural transition in a mainstream energy culture that has not assimilated the technologies yet (like Trinidad). Featured and expected changes to a regime's energy culture integrating SWHs are:

- The SWH becomes material culture.
- Adopting the SWH will be a practice; it is a compound practice made up of others such as gathering information and paying for the system.

- Flipping the electrical backup switch when needed will be a practice.
- Consuming solar energy will be a norm.
- Increased energy consumption and costs from specifically flipping the backup switch, and vice versa will be norms.
- Reduced energy consumption and costs from using solar water heating will be norms.
- The retail market, government policies, sunlight, cloud cover and climate change will be examples of external influences. Further, some external influences may remain the same, but their influence will be different, e.g. the sun may affect drying clothing outside currently, but it will also affect the heating efficacy of SWHs.

At the systems-level, SWHs can be thought of as a gateway technology for PV. Barbados has made significant strides in PV policies and installations. The island has utility-scale and household PV systems currently and so marks the beginning of a transition from solar water heating to/through solar PV – and specifically so distributed residential PV. These distributed systems have the support of policies such as import duty exemptions and tax credits. But the main driver is the RER which allows owners of such PV systems to sell power to the BL&P.

However, household PV has arguably not been mainstreamed just yet on the island. The last case study, O'ahu, has mainstreamed both household SWHs and PV and so Chapter 7 picks up where this one ends on PV and policies to illustrate how policies can accelerate and drive the mainstreaming of household PV.

References

BL&P (Barbados Light and Power Company Limited). (2014). *A Guide to Understanding Your Renewable Energy Rider (RER) Contract*. Retrieved from www.BL&Pc.com.bb/images/brochures/UNDERSTANDING%20YOUR%20RER%20CONTRACT%20(2).pdf.

BL&P (Barbados Light and Power Company Limited). (2015a). *How We Generate Electricity*. Corporate Information. Retrieved from www.BL&Pc.com.bb/co-mis/how-we-make-electricity.html.

BL&P (Barbados Light and Power Company Limited). (2015b). *Estimating Your Bill*. Understanding Your Bill. Retrieved from www.blpc.com.bb/bill-un/bill-est.html.

BL&P (Barbados Light and Power Company Limited). (2015c). *Fuel Clause Adjustment*. FAQs. Retrieved from www.blpc.com.bb/que/fuel-clause-adjustment.html.

BL&P (Barbados Light and Power Company Limited). (2015d). *Energy Riders*. Residential Services. Retrieved from www.BL&Pc.com.bb/cus-req/energy-riders.html.

BloombergNEF. (2018). Barbados Net Metering. *Climatescope*. Retrieved from http://global-climatescope.org/policies/1898.

Buchinger, J., Ince, D., Perch, L. & Hatvan, B. (2018). Supply(ier) side analysis – 5.3.1 PV. In *Barbados Sustainable Energy Industry Market Assessment Report* (pp. 24–28). Retrieved from www.ccreee.org/sites/default/files/project/files/barbados_market_assessment_report_-_final_2018-03-19.pdf.

148 Transitioning through household solar

Bugler, W. (2012). *Seizing the Sunshine: Barbados' Thriving Solar Water Heater Industry.* Inside Stories on Climate Compatible Development, Climate & Knowledge Development Network and Acclimatise. Retrieved from https://cdkn.org/resource/cdkn-inside-story-seizing-the-sunshine-barbados-thriving-solar-water-heater-industry/?loclang=en_gb.

Castalia Limited. (2010). *Sustainable Energy Framework for Barbados ATN/OC-11473-BA. Final Report-Volume 1.* Government of Barbados and Inter-American Development Bank. Retrieved from https://bajan.files.wordpress.com/2011/07/barbados-sustainable-energy-framework-vol-i.pdf.

CDB (Caribbean Development Bank). (2014). *A New Paradigm for Caribbean Development: Transitioning to a Green Economy.* Barbados. Retrieved from www.caribank.org/publications-and-resources/resource-library/thematic-papers/study-new-paradigm-caribbean-development-transitioning-green-economy.

CEIS (Caribbean Energy Information System). (2015, November 2015). *Solar Swells in Barbados; Capacity Set to Double in 2016.* Retrieved from www.ceis-caribenergy.org/solar-swells-in-barbados-capacity-set-to-double-in-2016-2/.

Cressman, D. (2009). *A Brief Overview of Actor-Network Theory: Punctualization, Heterogeneous Engineering & Translation.* British Columbia, Canada: ACT Lab/Centre for Policy Research on Science & Technology (CPROST), School of Communication, Simon Fraser University. Retrieved from https://summit.sfu.ca/item/13593.

Curry, B. (2007). *Solar Hot Water Systems.* Copper Initiative. Retrieved from http://copperplumbing.org.uk/sites/default/files/content_attachments/solar-hot-water-systems_2.pdf.

Del Chiaro, B. & Telleen-Lawton, T. (2007). *Solar Water Heating: How California Can Reduce Its Dependence on Natural Gas.* California: Environment California Research and Policy Center. Retrieved from http://area-net.org/wp-content/uploads/2016/01/Solar-Water-Heating_California.pdf.

ECLAC (Economic Commission for Latin America and the Caribbean). (2003). Solar Technology and Sustainable Development: Building on the Solar Dynamics Experience: Barbados. In *Section 2.4(15) – Science and Technology, Volume 2 – Small Island Developing States*, Santiago, Chile. Retrieved www.cepal.org/cgi-bin/getprod.asp?xml=/iyd/noticias/paginas/5/32105/P32105.xml&xsl=/iyd/tpl/p18f.xsl&base=/iyd/tpl/top-bottom.xsl.

FTC (Fair Trading Commission). (2016). *Decision on the Motion to Review the Renewable Energy Rider (RER).* Retrieved from www.ftc.gov.bb/index.php?option=com_content&task=view&id=309.

Gischler, C., Medina, R., Ordóñez, J., Echevarría, C., Alleng, G., Beall, E., Medeazza, G. M. V., Franklin, R. & Buchara, D. (2009). *Sustainable Energy Framework for Barbados (BA-T1007) Plan of Operations.* Inter-American Development Bank. Retrieved from www.energy.gov.bb/web/sustainable-energy-framework-for-barbados.

GoB (Government of Barbados). (2015). *National Sustainable Energy Policy (Revised).* Retrieved from www.energy.gov.bb/web/component/docman/doc_download/58-nsep.

Gordon, W. (2011). Behavioural Economics and Qualitative Research – A Marriage made in Heaven? International *Journal of Market Research*, 53(2): 171–185. Retrieved from https://doi.org/10.2501%2FIJMR-53-2-171-186.

Goss, B. (2010). *Choosing Solar Electricity: A Guide to Photovoltaic Systems*. Machynlleth, UK: Centre for Alternative Technology.

Gray, F., Koo, J., Chessin, E., Rickerson, W. & Curti, J. (2015). *Solar Water Heating Techscope Market Readiness Assessment. Reports for: Aruba, Bahamas, Barbados, Dominican Republic, Grenada, Jamaica, St. Lucia, Trinidad and Tobago*. Paris, France: Division of Technology, Industry and Economics, Global Solar Water Heating Initiative, United Nations Environment Programme. Retrieved from www.solarthermalworld.org/sites/gstec/files/story/2015-10-16/activity_5_-_swh_techscope_assessments_of_eight_caribbean_countries_report.pdf?ref_site=ppiaf.

Headley, O. St. C. (1998). Solar Thermal Application in the West Indies. *Renewable Energy*, 15(1–4): 257–263. Retrieved from https://doi.org/10.1016/S0960-1481(98)00170-0.

Hill, F., Lynch, H. & Levermore, G. (2011). Consumer Impacts on Dividends from Solar Water Heating. *Energy Efficiency*, 4(1): 1–8. Retrieved from https://doi.org/10.1007/s12053-010-9086-2.

Hohmeyer, O. (2015). *A 100% Renewable Barbados and Lower Energy Bills. A Plan to Change Barbados' Power Supply to 100% Renewables and its Possible Benefits*. Flensburg, Germany: Center for Sustainable Energy Systems, Flensburg University of Applied Sciences. Retrieved from www.uni-flensburg.de/fileadmin/content/abteilungen/industrial/dokumente/downloads/veroeffentlichungen/diskussionsbeitraege/znes-discussionspapers-005-barbados.pdf.

Husbands, J. (2010). *The Financial Benefits of Solar Hot Water Systems to Barbados*. Presentation published by Solar Dynamics Limited. Retrieved from www.solartherm alworld.org/sites/gstec/files/2.Mr_.James_Husbands.pdf.

Husbands, J. (2016). *The History and Development of the Solar Hot Water Industry in Barbados*. Presentation published by Solar Dynamics Limited. Retrieved from http://solardynamicslimited.com/wp-content/uploads/2016/10/Histor-Solar-Water-Heating-Industry-Barbados.pdf.

IEA-ETSAP and IRENA (International Energy Agency Energy Technology Systems Analysis Programme and International Renewable Energy Agency). (2015). *Solar Heating and Cooling for Residential Applications – Technology Brief*. Retrieved from www.solarthermalworld.org/sites/gstec/files/news/file/2015-02-27/irena-solar-heating-and-cooling-2015.pdf.

Ince, D. (2017). *Final Draft of the Energy Policy (2017–2037)*. Presented to the Division of Energy and Telecommunications, Prime Minister's Office, Government of Barbados, Barbados. Retrieved from www.energy.gov.bb/web/component/docman/doc_download/76-final-draft-of-national-energy-policy.

Ince, D. (2018). *Barbados National Energy Policy (2017–2037)*. Presented to the Division of Energy and Telecommunications, Prime Minister's Office, Government of Barbados, Barbados. Retrieved from www.energy.gov.bb/web/component/docman/doc_download/86-barbados-national-energy-policy-2017-2037.

Jensen, T. L. (2000). *Renewable Energy on Small Islands* (2nd ed.). Forum for Energy and Development. Retrieved from www.gdrc.org/oceans/Small-Islands-II.pdf.

Kakaza, M. & Folly, K. A. (2015). Effect of Solar Water Heating System in Reducing Household Energy Consumption. *IFAC-PapersOnLine*, 48(30): 468–472. Retrieved from https://doi.org/10.1016/j.ifacol.2015.12.423.

Labay, D. G. & Kinnear, T. C. (1981). Exploring the Consumer Decision Process in the Adoption of Solar Energy Systems. *Journal of Consumer Research*, 8(3): 271–278. Retrieved from www.jstor.org/stable/2488885.

Madden, M. (2019, January 24). RER review needed. *Barbados Today*. Retrieved from https://barbadostoday.bb/2019/01/24/rer-review-needed/.

Manning, R. A. (2015). *Renewable Energy's Coming of Age: A Disruptive Technology?* Washington, DC: Brent Scowcroft Center on International Security and Global Energy Center, Atlantic Council. Retrieved from www.files.ethz.ch/isn/195316/Renewable_Energy.pdf.

Mark, D. (2016). [*Untitled illustration of teapots on a cooking stove*]. Pixabay. Retrieved from https://pixabay.com/photos/teapots-pots-cook-stove-flame-1858601/.

Marzolf, N. C., Cañeque, F. C. & Loy, D. (2015). *A Unique Approach for Sustainable Energy in Trinidad and Tobago*. Washington, DC: Inter-American Development Bank. Retrieved from www.energy.gov.tt/wp-content/uploads/2016/08/A-Unique-Approach-for-Sustainable-Energy-in-Trinidad-and-Tobago.pdf.

McIntyre A., El-Ashram, A., Ronci, M., Reynaud, J., Che, N., Wang, K., Acevedo, S., Lutz, M., Strodel, F., Osueke, A. & Yun, H. (2016). *Caribbean Energy: Macro-related Challenges*. IMF Working Paper, International Monetary Fund. Retrieved from www.imf.org/external/pubs/ft/wp/2016/wp1653.pdf.

McVeigh, J. C. 1(984). *Energy Around I World: An Introduction to Energy Studies. Global Resources, Needs, Utilization*. Oxford, UK: Pergamon Press Limited.

Milton, S. & Kaufman, S. (2005). *Solar Water Heating as a Climate Protection Stratelthe Role for Carbon Finance*. Massachusetts: Green Markets International Incorporated. Retrieved from www.green-markets.org/Downloads/SWH_carbon.pdf.

Moore, W., Alleyne, F., Alleyne, Y., Blackman, K., Blenman, C., Carter, S., Cashman, A., Cumberbatch, J., Downes, A., Hoyte, H., Mahon, R., Mamingi, N., McConney, P., Pena, M., Roberts, S., Rogers, T., Sealy, S., Sinckler, T. & Singh, A. (2014). *Barbados' Green Economy Scoping Study*. Barbados: Government of Barbados, University of West Indies – Cave Hill Campus, and United Nations Environment Programme. Retrieved from www.un-page.org/file/1593/download?token=uiUKfl0J.

Namias, O. (2008). The Hidden Dimensions of Electrical Architecture. In M. Rüdiger (Ed.), *The Culture of Energy* (pp. 93–102). Newcastle: Cambridge Scholars Publishing.

Niu, S., Hong, Z., Qiang, W., Shi, Y., Liang, M. & Li, Z. (2017). Assessing the Potential and Benefits of Domestic Solar Water Heating System Based on Field Survey. *Environmental Progress and Sustainable Energy*, 37(5): 1781–1791. Retrieved from https://doi.org/10.1002/ep.12827.

NREL (National Renewable Energy Laboratory). (1996). *Solar Water Heating*. United States Department of Energy. Retrieved from www.nrel.gov/docs/legosti/fy96/17459.pdf.

NREL (National Renewable Energy Laboratory). (2015a). *Playbook Lesson Learned. Phase 3 Project Preparation. Solar Hot Water Heater Industry in Barbados*. Energy Transitions Initiative-islands. United States Department of Energy. Retrieved from https://energy.gov/sites/prod/files/2015/03/f20/phase3-barbados.pdf.

NREL (National Renewable Energy Laboratory). (2015b). *Energy Snapshot: Barbados*. Energy Transition Initiative-Islands, United States Department of Energy. Retrieved from www.nrel.gov/docs/fy15osti/64118.pdf.

Perlack, B. & Hinds, W. (2003). *Evaluation of the Barbados Solar Water Heating Experience.* Retrieved from http://solardynamicsltd.com/wp-content/uploads/2010/07/SWH-report1-2.pdf.

READ (Research and Economic Analysis Division). (2014). *State of Hawaii Electricity Generation and Consumption in 2013 and Recent Trends.* State of Hawaii: Hawaii Economic Issues. Department of Business, Economic Development and Tourism. Retrieved from http://files.hawaii.gov/dbedt/economic/data_reports/reports-studies/ElectricityTrendsReport2014.pdf.

Richerson, P. J., Mulder, M. B. & B. Vila, B. (2001). *Diffusion of Innovations*. In *Principles of Human Ecology* (pp. 350–364). California: University of California, Davis. Retrieved from www.des.ucdavis.edu/faculty/Richerson/BooksOnline/101text.htm.

Rogers, E. M. (1983). The Innovation-Decision Process. In *Diffusion of Innovations* (3rd ed.) (pp. 163–209). New York: The Free Press.

Rogers, T. (2016). *Planning for Renewable Energy: A 100% Renewable Energy Vision for Barbados.* Presentation for the Barbados Town Planning Society, the University of the West Indies, Cave Hill, Barbados. Retrieved from www.barbadosplanningsociety.org/wp-content/uploads/Tom-Rogers-100-percent_Sept-2016.pdf.

Rogers, T., Chmutina, K. & Moseley, L. L. (2012). The Potential of PV Installations in SIDS – An Example in the Island of Barbados. *Management of Environmental Quality*, 23(3): 284–290. Retrieved from https://doi.org/10.1108/14777831211217486.

Samuel, H. A. (2013). *A Review of the Status of the Interconnection of Distributed Renewables to the Grid in CARICOM Countries.* Castries, St. Lucia: Caribbean Renewable Energy Development Programme, Caribbean Community and Deutsche Gesellschaft für Internationale Zusammenarbeit (GIZ) GmbH Germany. Retrieved from www.scribd.com/document/252916035/CREDP-GIZ-A-Review-of-the-Status-of-the-Interconnection-of-Distributed-Renewables-to-the-Grid-in-CARICOM-countries-2013.

Schelly, C. (2014). Residential Solar Electricity Adoption: What Motivates, and What Matters? A Case Study of Early Adopters. *Energy Research and Social Science*, 2: 183–191. Retrieved from https://doi.org/10.1016/j.erss.2014.01.001.

SEforALL (Sustainable Energy for All). (2012). *Barbados: Rapid Assessment and Gap Analysis.* Retrieved from https://seforall.org/sites/default/files/Barbados_RAGA_EN.pdf.

Sepp, S. (2014). Liquefied Petroleum Gas. In H. Volkmer (Ed.), *Multiple-Household Fuel Use – A Balanced Choice Between Firewood, Charcoal and LPG* (pp. 33–41). Eschborn, Germany: Deutsche Gesellschaft für Internationale Zusammenarbeit (GIZ) GmbH on behalf of the Federal Ministry for Economic Cooperation and Development (BMZ). Retrieved from www.cleancookingalliance.org/resources/287.html.

Sovacool, B. K. (2009). The Cultural Barriers to Renewable Energy and Energy Efficiency in the United States. *Technology in Society*, 31(4): 365–373. Retrieved from https://doi.org/10.1016/j.techsoc.2009.10.009.

Stephenson, J. (2012). *Energy Cultures: The Concepts and its Applications (So Far).* Presentation given at University College London, London.

Thompson, E. H. (2015). *Lessons from the Reform of the Barbados Energy Sector*. Presentation published by the Energy Sector Management Assistance Program for Small Island Developing States, The World Bank, Washington, DC. Retrieved from https://esmap.org/sites/esmap.org/files/DocumentLibrary/3a%20-%20Elizabeth%20WB%20ESMAP%20PP_Optimized.pdf.

UN (United Nations). (2002). Barbados Country Profile. Johannes Summit 2002 Country Profile Series. Retrieved from www.un.org/esa/agenda21/natlinfo/wssd/barbados.pdf.

Wallenborn, G. & Wilhite, H. (2014). Rethinking Embodied Knowledge and Household Consumption. *Energy Research and Social Science*, 1: 56–64. Retrieved from https://doi.org/10.1016/j.erss.2014.03.009.

7 Solar water heating, PV and policy implementation

Introduction

The last chapter showed that Barbados is in the process of transitioning to/through distributed PV after having mainstreamed household SWHs. This chapter builds on this since Oʻahu has mainstreamed household PV, so it is part of the island's residential electricity regime.

Like Barbados, Hawaiʻi's SWH industry also had its beginnings in the 1970s and this also includes the associated policy support. For example, in 1976/1977 tax credits for SWHs were implemented (Yailen et al., 2012, 341; InSynergy Engineering, 2012, 1; Haleakala Solar, 2014). These credits allow system owners to get an additional refund with their annual tax return for installing an eligible solar system.

These incentives markedly reduced the costs of systems for households, and at the time, there were nearly 50,000 systems installed which meant that one in every three single-family homes had a SWH (Haleakala Solar, 2014). Since then, other initiatives which have supported SWHs include solar rebates instituted in 1996 and legislation passed in 2010 which mandates new single-family homes to have a SWH installed (Yailen et al., 2012; Haleakala Solar, 2014).

However, the tax credits programme was arguably the most effective at encouraging SWH adoption – to the point where there has been a direct historical correlation between the availability of tax credits and the adoption SWHs in Hawaiʻi (InSynergy Engineering, 2012). In 1977 there was a 10% State tax credit available; 40% (10% State plus 30% Federal) in 1978/1979; 50% (10% State plus 40% Federal) between 1980 to 1985; 15% State credit only between 1986 to 1988; 20% State credit in 1989; 35% State credit from 1990 to 2005; and then 65% (35% State plus 30% Federal) credit between 2006 to 2011 (InSynergy Engineering, 2012).

Based on the NCCETC's policy database, this last credit level is still available (at least as recent as 2018) though the impression given from the policy officials interviewed in Oʻahu in 2016 was that the programme's credit levels may be revised to reflect changing market conditions.

Nevertheless, since the 1970s, 103,305 SWHs have been installed in the State between then and 2011 (InSynergy Engineering, 2012). InSynergy

Engineering caveat that the number in operation is likely to be lower because it accounts for replacement systems as well. Today, there are perhaps around 90,000 systems and the ratio has changed to one in every four single-family homes owning a SWH (HECO, 2019a). This means that almost 75% of homes do not have solar hot water (Yailen et al., 2012, 346). Nevertheless, these sorts of figures have ranked the State first with respect to SWHs installed per capita compared to the rest of the US (Lim, 2011, 21; HECO, 2019a).

The policy officials' views from the interviews conducted converged on the belief that the mid- to late 2000s, that is, roughly the 2008 to 2012 period, was a key time for Hawai'i's solar energy growth because there were developments which encouraged its widespread adoption (particularly so PV). There were several specific factors which supported solar as an alternative and understanding the factors as landscape and regime-level forces will put them into the context of the theory from Chapter 2 and illustrate the forces influencing the transition to/through PV.

At the landscape level, the factors were the global recession and market recovery, together with its associated increases in the international oil/petroleum prices. Additionally, the plummet in the international cost of solar systems was another. The price drop was a result of the introduction of dedicated silicon feedstock for manufacturing panels (versus the prior use of waste from the semi-conductor industry), as well as the mass production of cheap Chinese brands adopted in European and North American markets.

At the regime level, the influence of the domestic electricity costs on the island as well as the implementation of policies which supported PV drove its mainstreaming. In the case of the last, policies such as a FiT, NEM, tax credits (like those for SWHs) and property tax exemptions were made accessible around this time (PUC, 2010; NCCETC, 2018a, 2018b, 2018c; R. P. Delio and Company, 2012).

The landscape and regime-level factors which emerged from the policy officials' views show that the mainstreaming of PV and solar energy in O'ahu (and by extension Hawai'i) was not an endogenous process or solely due to one causative agent. So, having given this brief insight into O'ahu and Hawai'i's solar hot water transition to PV, now the incumbent PV-influenced residential electricity regime can be characterized.

Characterizing O'ahu's incumbent residential electricity regime

Energy sourcing

Hawai'i has no indigenous fossil fuels (GEC, 2006; IER, 2013; Alm, 2015). Until recently, the State was almost 90% dependent on imported oil/petroleum (RMI, 2008, 5; Hunter, Westlake & Griffith, 2013, 14; HNEI, 2016, 5). But through the implementation of local renewable energy technologies, energy efficiency interventions, policies and legislative directives, it is closer to 80% (State of Hawaii, 2019a).

Most of the oil/petroleum imports come from the Middle East (GEC, 2006; RMI, 2008) – namely Saudi Arabia and Oman, but also further afield from places like Indonesia and Malaysia (Shupe, 1982; Jensen, 2000; HECO, 2011). Each barrel of oil imported into the State produces jet fuel, diesel, gas, gasoline, marine fuels, as well as bunker-type oil (Alm, 2015).

The annual petroleum import bill ranges from US$4.5 to 6 billion (GEC, 2006, 4; Baker, 2014; Morita, 2014; Pintz & Morita, 2017). As an oil importer, oil price decreases benefit Hawai'i because of lower importation costs, and vice versa when prices rise (RMI, 2008; Johnson & Chertow, 2009; Alm, 2015) – similarly to Barbados. Therefore, the State is an energy price-taker (Bass & Dalal-Clayton, 1995; RMI, 2008). It is also the most petroleum-dependent State in the U.S. (Jensen, 2000; GEC, 2006; Corbus et al. 2013; DBEDT, 2017; DBEDT, 2019).

Hawai'i's electricity sector uses an estimated 24.5% of the State's fossil fuel imports (DBEDT, 2019, 4). Data from the DBEDT (2019, 2) show that Hawai'i's power production mix is largely made up of imported petroleum (61.3%) and coal (11.9%). However, there are also smaller contributions from renewable resources such as small-scale solar (9.3%), wind (4.9%), geothermal (2.9%), biomass (2.8%), utility-scale solar (1.9%) and hydro (0.9%) (DBEDT, 2019, 2).

Power generation

Each Hawai'ian island has its own electrical grid (GEC, 2006; Kaya & Yalcintas, 2010; R. P. Delio and Company, 2012; Piwko et al., 2012). There are six independent grids that transmit and distribute electricity across the archipelago's eight main islands (Jensen, 2000; RMI, 2008; Morita, 2014).

There are four electric utilities: the Kauai Island Utility Cooperative on Kaua'i; the HECO servicing O'ahu; a HECO subsidiary, the HELCO, on Hawai'i Island; and another HECO subsidiary, the MECO, serving the islands of Maui, Moloka'i and L'na'i (HNEI, 2008; Kaya & Yalcintas, 2010; Morita, 2014; DBEDT, 2014; Gorak, 2016). The last three fall under the HEI.

It is O'ahu's grid and the HECO that are of interest, however. O'ahu's total power generation capacity is rated at 2,421.7 MW and made up of roughly 1,320 MW from utility-owned generation and 1,101.7 MW from IPPs (HECO, 2019b) (see Table 7.1). The HECO therefore generates its own power as well as purchases power from IPPs (RMI, 2008; DBEDT, 2014; DBEDT, 2017).

As shown in the last section, Hawai'i has several renewable energy resources in its power generation mix in addition to the centralized fossil fuel-powered power generation units. In that regard and relative to the scope of the book, solar energy has become the main renewable energy resource in the State as shown by the DBEDT (2017) and DBEDT (2019).

O'ahu has utility-scale PV as well as distributed PV. There are around 16 large-scale solar plants in O'ahu and examples include the 28-MW EE

Table 7.1 Showing Oʻahu's power generation capacity

Unit	Capacity (MW)	Ownership
Waiau (oil)	500	HECO
Kahe (oil)	650	
Campbell Industrial Park (biofuel)	120	
Schofield (biofuel/diesel)	50	
TOTAL	1,320	
HPOWER (waste-to-energy)	68.5	IPPs
Kalaeloa Partners (oil)	208	
AES-Hawaiʻi (coal)	180	
Airport Emergency Facility (biofuel)	8	
Kahuku Wind	30	
Kawailoa Wind	69	
Waiʻanae Solar	27.6	
Par Hawaiʻi	18.5	
Island Energy Service	9.6	
Waihonu Solar	6.5	
Aloha Solar Fund 1	5	
Kalaeloa Solar Two	5	
Kalaeloa Renewable Energy Park (PV)	5	
Kapolei Sustainable Energy Park (PV)	1	
Customer-sited solar	460	
Total	1,101.7	

Source: Compiled by the Author using figures from the HECO (2019b).

Waianae Solar plant (the largest) and the 245-kW Kahumana PV project (the smallest) (DBEDT, 2019, 28) – and the DBEDT (2019) lists 14 more projects in the pipeline. However, unlike this utility-scale solar, distributed PV is characterized by decentralized, smaller generators (Johnson & Chertow, 2009; Goss, 2010) – as is the case for household PV. Therefore, with household PV, the technical linkage between electricity generation and consumption is shorter compared to utility-scale PV because the latter connects to the grid at the transmission level whilst distributed PV connects at the distribution level (Codiga, 2013).

However, distributed PV dominates the Hawaiʻian market (DBEDT, 2017, 18). For example, of Oʻahu's total renewable energy generation in 2018 of 2,520.2 GWh, solar energy accounted for 48.2% of this (1,215.1 GWh), and distributed PV specifically contributed 39.6% (998.1 GWh) (DBEDT, 2019; 14). The HECO (2019c) also state that 96% of distributed PV systems are residential.

Coffman, Allen and Wee (2018; 13) show that the island's penetration of residential PV ranges from none to over 40% – where penetration here refers to the proportion of occupied housing units with PV installed. This is particularly useful to consider in light of the theoretical 16% adoption pseudo-threshold linked to mainstreaming from Chapter 2. To put this into more explicit context, there are well over 60 districts in Oʻahu with over 21% adoption based on a map produced in Coffman, Allen and Wee's work.

Nevertheless, the changes observed due to distributed PVs' mainstreaming mean that the relationship between the utility, energy producer and consumer can be quite different compared to the pre-existing design (R. P. Delio and Company, 2012); an electricity system that has distributed PV integrated into it is more complex (Wüstenhagen, Wolsink & Burer, 2007; ABB, 2012; R. P. Delio and Company, 2012; IEC, 2012; WEC, 2016), as is consistent with the notion that each energy transition increases the regime's complexity (Chabrol, 2016).

For example, in the case of distributed PV here, the increased complexity results from the two-way streams of electricity where solar electricity is exported from households and electricity is imported by households (Brooks, 2003). Further, as is the case with several of the policies mentioned in the Introduction, for example NEM and FiTs, this two-way exchange also applies to information and finance since households are either paid for their solar electricity exported to the grid and/or pay the utility for consuming its electricity from the grid based on the corresponding billing/crediting information.

Electricity usage

Hawaiʻi's dependence on imported oil/petroleum exposes its energy sector to the international markets' fluctuant prices and these fluctuations are reflected in the electricity rates (READ, 2014; Morita, 2014; Hollier, 2015; Pintz & Morita, 2017). For households this means that when international oil prices are high, the retail electricity rate is high, and vice versa, because the importation costs are transferred onto the local tariffs through the ECAC which allows the risks associated with fluctuant oil/petroleum prices to be passed directly onto consumers (HECO, 2011; Pintz & Morita, 2017). Figure 7.1 illustrates these dynamics in Oʻahu but the mechanism is similar to the FCA highlighted in Chapter 6 for Barbados.

The READ (2014, 33) and Morita (2014) outline that 60 to 70% of Hawaiʻi's electricity costs are due to the fuel and purchased power costs as well as their related taxes. For example, the READ (2014; 43) shows that the residential rate in 1990 was roughly US$10.26 cents per kWh and rose to just over 37.34 cents by 2012 largely because of the fuel and purchased power costs.

Morita (2014; 10) also shows another example through the per kWh breakdown of the cost recovery for a 'typical' customer's bill in Oʻahu based on a rate of 30.9 cents per kWh. She shows that the costs of the fuel (12.4 cents); purchased power (7.7 cents); operations and maintenance (4.1 cents); revenue tax (2.9 cents); return and other (1.6 cents); depreciation (1.5 cents); and income tax (0.7 cents) are all recovered but the fuel and purchased power costs make up the bulk of the costs.

158 Transitioning through household solar

Landscape influences on the markets' prices

Landscape factors: Fossil fuel reserve projections, Production, U.S. Shale oil, Geopolitical developments, Demand, Non-OECD demand, Legislation, Financial markets, Value of the U.S. dollar, OPEC quotas

Oil imports (>80% of Hawai'i's energy) (Mainly from the Middle East)

Decrease in market price / **Increase in market price**

Landscape	Landscape
Regime	Regime

Importation bill (Billions annually)

Lower import bill — Technically limited grid, Islandic geography; The Hawai'ian Electric Company sends the new tariff application to be vetted and approved by the Public Utilities Commission. However, the Energy Cost Adjustment Clause enables the transfer of the oil price onto the electricity rate.

Higher import bill — Natural energy sources available, Isolated grid, Aggressive clean energy policy, Takes 2 to 3 months to be reflected in the tariff.

Electricity generation, transmission and distribution costs

Cost Recovery

Lower tariffs — Hawai'i's dependence on oil imports makes it a price-taker and limits its energy autarky.

Consumer electricity costs

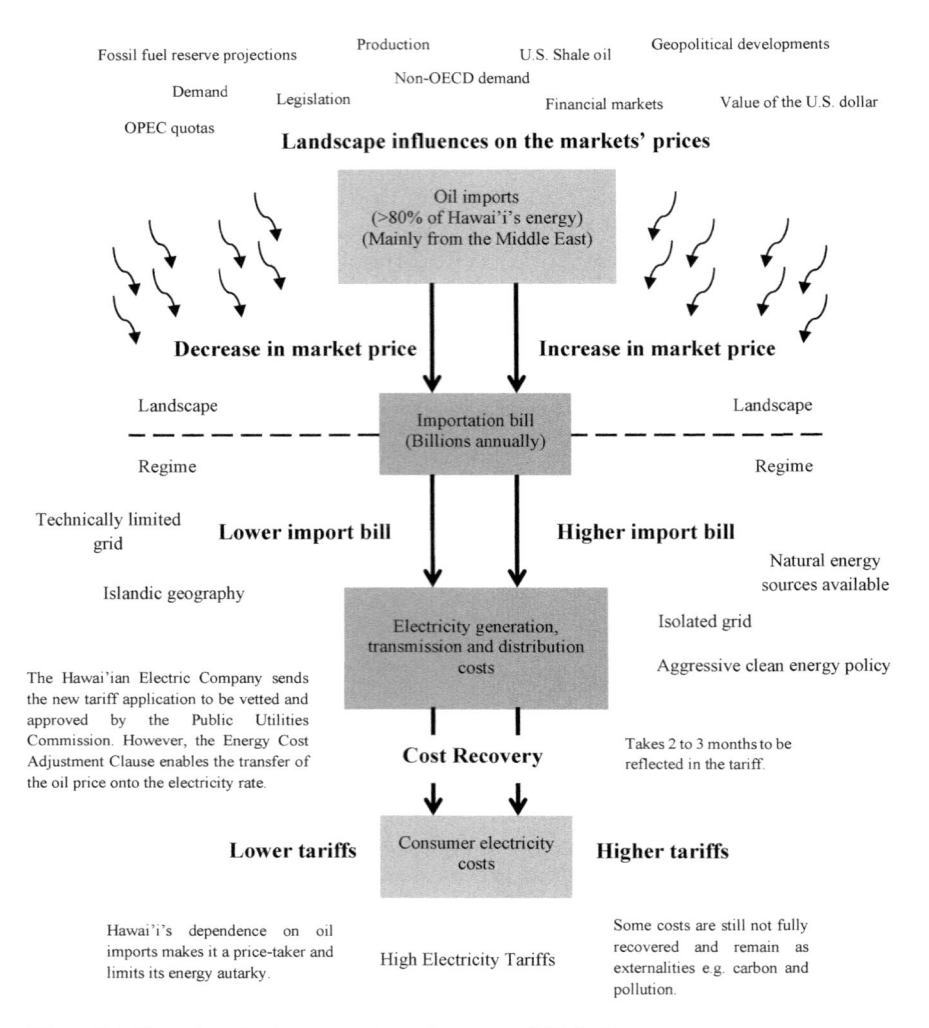

Higher tariffs — Some costs are still not fully recovered and remain as externalities e.g. carbon and pollution.

High Electricity Tariffs

Figure 7.1 Showing the import-oriented nature of O'ahu's energy system.

The average retail rate in Hawai'i is usually between US$30 and 37 cents per kWh, and in O'ahu, between 26 cents and 32 cents (Sillaman, 2015; Electricity Local, 2019; DBEDT, 2017, 3; DBEDT, 2019, 5). The State's retail rates are often twice that of the national average (Sillaman, 2015; DBEDT, 2017; DBEDT, 2019) and are the highest in the country (Stockton, 2004; GEC, 2006; Piwko et al., 2012; Sillaman, 2015; Electricity Local, 2019).

However, despite the energy transition through grid-connected household PV, the relationships between the utility and households in O'ahu still revolve around the technical distribution of electricity, its consumption and the payments for it. Data from the DBEDT (2019, 5) show that the average household energy consumption has decreased from 584 kWh per month in 2011 to 494 kWh in 2018 for Hawai'i, and from 609 kWh to 493 kWh in O'ahu within the same period.

The State's average monthly bill also decreased from US$202 to $163, and in O'ahu, from $195 to $155 over this timeframe (DBEDT, 2019). Other data from the DBEDT (2017) supports these trends by showing that the kWh sold to the residential sector has been decreasing. These trends are a result of household renewable energy like PV and energy efficiency interventions though Coffman. Aleen and Wee (2015) point out that it is mostly because of efficiency gains.

Relative to household PV, the HSEO (2014) points out that solar system installations require permit applications and approvals, and householders need to apply for a permit from the HECO as well as DPP – the latter is involved because installing solar is a construction-related project. So, given the significance of distributed PV in the last section, it is unsurprising that most of these permit applications are residential (READ, 2015).

In terms of households acquiring solar systems, the most obvious interaction is between households and solar retailers where householders purchase the systems from retailers. But based on the above permit application interactions it can be expected that retailers receive guidance from the HECO and DPP on issues such as codes of practice, health and safety, certification and registration. The DBEDT (2003) states that solar system dealers and professional contractors must also have a solar speciality license to do installations whilst in the case of do-it-yourself projects, the tradespersons involved must be licensed.

Electricity regulation and governance

Given the appreciation for the technical electricity generation-consumption pathway in the preceding sections, this one will look at the roles of the wider institutions involved in regulating electricity as a commodity in O'ahu; the roles of the PUC and DCA, HSEO, HSL, Office of the Governor, the UofH and HNEI, the Blue Planet Foundation and HSEA will be briefly elaborated on as contextual examples.

The PUC is one of the institutions responsible for reducing the State's reliance on fossil fuels and promoting increased renewable energy generation (Marumoto, 2013). It is the State's independent utilities regulator and the interviews with policy officials revealed that the PUC regulates the HECO and not the IPPs. The DCA is an institution that is related to the PUC because it is a State-established entity set up with the purpose of protecting and representing consumer-interests before the PUC as it relates to electricity rate-increase applications from the HECO at public hearings (State of Hawaii, 2019b).

The HSEO works on solutions geared towards removing the barriers to renewable energy and energy efficiency rollouts, aligning regulations and

160   *Transitioning through household solar*

legislations with clean energy objectives and promoting Hawai'i-based clean energy research and investments (Marumoto, 2013). It seeks to lead the transition towards energy independence by diversifying the energy portfolio; encouraging market competition; connecting the Hawai'ian grids; promoting Hawai'i as an energy innovation testbed; and balancing the considerations of stakeholders (Marumoto, 2013).

Further, the HSEO's role is set within the State's four energy objectives of fostering: 1) reliable and effective energy systems capable of supporting society's needs; 2) increased energy independence where the ratio of indigenous to imported energy use is increased; 3) increased energy security in light of any threats to Hawai'i's energy system; and 4) the reduction of greenhouse gas emissions from energy supply and usage (HNEI, 2008).

The HSL enacts policies and this includes those related to energy and electricity. The policy officials' interviews framed the Legislature's laws as items that are developed with the input of the stakeholders involved in Hawai'ian energy and once they become enshrined in law they serve as directives for all actors with specific relevance for the electricity regulators and market, for example PUC, HECO and HSEO. Further, the interviews showed that there is no exclusive top-down or bottom-up relationship between the PUC and HSEO during policymaking.

Transitioning to renewables involves a multiplicity of actors and interests (Gorak, 2016) and making policies are compromises between stakeholders' perspectives and achieving consensus (Beerepoot & Beerepoot, 2007) – and so, the interviews acknowledged that one of the legislative challenges, if not the greatest, is reconciling different actors' views.

The legislature drafts and designs legislation but the power of ratification resides with the Governor (Marumoto, 2013). The Governor is responsible for setting the broad goals of the State and the priorities for achieving these goals (Marumoto, 2013). Whilst the Governor is engaged in a wide range of governance issues and not just electricity or energy, the policy officials pointed out that the Governor has a direct and specific impact on the PUC through the appointment of Commissioners – and these Commissioners should not have any interest in or income from any of the State's utilities (Marumoto, 2013).

The HNEI is a research unit of the UofH which was established to work with the Government to reduce the State's fossil fuel dependence (HNEI, 2016). It has interests in alternative and renewable energy, building efficiency, transport, grid integration and electrochemical power systems (SOEST, 2013; HNEI, 2016). The legislature even established a statute that supports the Institute's collaboration with State and federal institutions (SOEST, 2013) – institutions like the PUC, HECO, the legislature itself, and HSEO (SOEST, 2013; HNEI, 2016). Therefore, the University and Institute are part of the governance through their supporting mandates.

The Blue Planet Foundation's objectives target policymakers as well as communities to influence the top-down and bottom-up segments of the energy system to inspire persons to believe in the possibility of transitioning away from fossil fuels, that is, a form of socio-environmental activism. The other NGO example being looked at is the HSEA. It is a professional trade association affiliated with the solar energy industry and is made up of businesses and firms that support solar energy along all segments of the supply chain (HSEA, 2012; Cole-Brooks, 2014). So, the Association adds an overarching industrial solar voice within the regime through legislative, regulatory and educational outreach initiatives for businesses and homes.

This overview of the institutional makeup of the residential electricity regime sets the tone for the rest of the chapter because it is at this point where the policies that have helped mainstream household PV can be looked at – policies coordinated by several of the institutions touched on here.

Policies as instruments of mainstreaming solar energy

Coffman et al. (2013), the EEEI (2014), Hunter, Westlake and Griffith (2013), GE Energy Consulting (2015) and the DBEDT (2015) state that RPSs, NEM, tax incentives, the FiT, green infrastructure bonds, rebates, loan programmes and the pay-as-you-save scheme were some key strategies that were implemented to mainstream solar energy in the State. But the impression given from the interviews with policy officials was that RPSs, NEM and tax incentives were/are the most influential (thus far). In that regard, RPSs, NEM (and FiTs for comparison) will be overviewed with some considerations for tax incentives where relevant in the interest of illustrating how such policy mechanisms support the mainstreaming of solar energy.

Renewable portfolio standards

RPSs have been around since the 1990s in the U.S. (Wiser et al., 2007); are the country's most common State-level policy (Cory et al., 2009); and are key drivers of the national renewable energy industry (Burnett, 2011; Zhai, 2013; PUC, 2013). The NREL (undated) notes that RPSs can be customized based on the utility company-ownership and technology preferences. So, the States in the U.S., utilizing RPSs would have enacted them based on their specific contexts (HNEI, 2008).

RPSs are regulatory instruments designed to stimulate an increase in renewable energy production by mandating electricity suppliers to incorporate renewables into their supply mixes and sales through specified numerical targets (Cory & Swezey, 2007; Cory et al., 2009; Lyon & Yin, 2010; Zhai, 2013; NREL, undated) – and these generation targets should increase over time (Wiser et al., 2008; Lyon & Yin, 2010; Yin & Powers, 2010).

In Hawai'i's case, the RPSs of '7% by 2003' evolved to '8% by 2005', '10% by 2010', '15% by 2015', '30% by 2020', '40% by 2030' and '70% by 2040' since its enactment in 2001 (RMI, 2008; Yalcintas & Kaya, 2009, 3268; PUC, 2013; Gorak, 2016, 7; PUC, 2018, 1). But linked to this last target is the HCEI.

The HCEI was launched in 2008 and is a programme that supports energy efficiency, the regulatory environment for clean energy and inter-island utility company collaborations to increase renewable energy generation and grid connections (State of Hawaii, 2019a; NREL, 2018; PUC, 2018). Braccio and Finch (2011) outline that the HCEI's four main strategies for the electricity sector are to deploy grid-compatible renewable energy technology; build confidence in renewables; explore the benefits of innovative and existing green technology; and align the electricity sector's regulatory and policy frameworks with cleaner energy.

The HCEI stipulated a goal of 70% clean energy by 2030 (PUC, 2018). However, this 70% is made up of the aforementioned '40% by 2030' RPSs target but with an added 30% of energy efficiency savings through the HCEI's EEPSs (Lim, 2011, 23; R. P. Delio and Company, 2012; Critz, Busche & Connors, 2013; PUC, 2013; EEEI, 2014). This shows how the HCEI's energy efficiency ambitions complement the RPSs in Hawai'i (NREL, 2018) – and having been implemented after the RPSs, Morita (2014) and Savenije (2015) believe that the EEPSs built on the RPSs' success over the years.

However, this is not to say that the EEPSs have not been effective. The EEPSs are like the RPSs in that they are both legislated directives. But the latter focuses on generation whilst EEPSs are focused on reducing demand and ensuring that the most benefit is derived from every kWh of electricity. A key programme-actor here is Hawaii Energy which is administered by Leidos Incorporated under a contract with the PUC as the Public Benefits Fee Administrator (PUC, 2018). Hawaii Energy is funded by ratepayers and specified amounts (as advised by the PUC) of these revenues are (re)invested into energy efficiency projects and improvements (PUC, 2018).

The State's progress in energy efficiency has been such that it remains a lower cost option compared to other clean energy options; Hawaii Energy's interventions will save nearly US$3.7 billion over the lifetime of the measures implemented thus far in addition to reducing peak demand by 150 MW, and the 2015 target of 1,375 GWh was surpassed by more than 50% with 2,030 GWh of savings for example (PUC, 2018, 1). The PUC (2018b) further outlines that the State is on track to meet the 2020 goals but also caveat that most of the cost-effective energy efficiency measures have already been implemented so there will need to be future investments to meet the State's clean energy ambitions.

Nevertheless, as with the EEPSs, the PUC coordinates with the HNEI to review the RPSs every five years (GEC, 2006, 6; PUC, 2018). Electricity suppliers also need to demonstrate their compliance annually (Wiser et al., 2008) so the utilities submit annual status reports (DBEDT, 2015) and the PUC reports on the status of the utilities' strategies to the State Legislature (PUC, 2013). To date there have been three RPSs status reports since the standards became legally binding (PUC, 2018, 4).

R. P. Delio and Company (2012) point out that Hawai'i's strong dependence on expensive fuel imports is one of the prime reasons why the State's RPSs have been so ambitious. However, the State of Hawaii (2019a) outlined that the State has now ratified a RPSs target of '100% by 2045', and it uses the previous ambits as interim targets (NREL, 2018, PUC, 2018); Hawai'i is the most populated archipelago with independent grids to have such an energy ambition (Renewables 100 Policy Institute, 2015).

The strongest RPSs use non-compliance penalties (through fines or alternative compliance payments) which are pre-determined amounts per kWh that a supplier pays if they fail to meet their RPSs and which cannot be recovered from ratepayers/customers (Cory & Swezey, 2007). The interviews conducted with policy officials in O'ahu suggested that the Hawai'ian utilities can be fined, and the PUC can ensure that some public benefit is derived from non-compliance, for example redirecting revenue into RPSs projects or redistributing it amongst ratepayers.

The penalties for not meeting the RPSs targets are consequently subject to the PUC's discretion and non-compliant utilities may be granted extensions to meet their obligations, or the penalty may be waived provided that non-compliance resulting from factors reasonably beyond a utility's control could be proven (HNEI, 2008; PUC, 2018) – and the PUC (2018b) also states that there are penalties and fines associated with not achieving EEPSs based on performance as well.

Relative to the solar focus of the book, a 2006 ruling enabled the PUC to prescribe which renewables should make up stipulated portions of electricity sales (HNEI, 2008; PUC, 2018, 4). In that regard, Hawai'i's four utilities meet the RPSs in their own ways (Cory et al., 2009) but Jensen (2000) and the HNEI (2008) note that the RPSs legislation also enables utilities to aggregate their renewable energy portfolios to meet their targets. Therefore, utilities can capitalize on their respective renewable energy resources to meet the overall corporate targets – and interestingly enough, PV is the largest contributor to the State's RPSs and specifically so distributed PV for the HEI's utilities as it makes up 9.9% of their current total renewable energy generation sales of 26.8% (PUC, 2018).

The most recent figures from the PUC (2018a, 6) show that the State's renewable energy generation amounted to 27.6% of the utilities' sales after having been at 8.9% a decade ago – and for reference, O'ahu's HECO is currently at 20.8% after having been at 4.3% in the same period (PUC, 2018, 6). The figures presented in the PUC's (2018a) RPSs report show that the Kauai Island Utility Cooperative has already achieved the 2020 and 2030 targets because they are currently at 44.5% and the HEI are currently at 26.8% and are 'highly likely' and 'likely' to meet the 2030 and 2040 targets, respectively. But the PUC (2018a) caveats that the 2040 and 2045 targets are filled with more uncertainty and there are many moving parts that need addressing. But despite this, the RPSs have been and still are effective drivers of Hawai'i's renewable energy growth (PUC, 2018).

Net energy metering

RPSs are broad policies and as hinted above, solar energy generation would be just a part of the renewable energy generation portfolio. One of Hawai'i's most impactful residential solar energy policies was its NEM programme. It had its roots as early as 1996 (Gorak, 2016), was instituted in 2001 (Morita & Mangelsdorf, 2015), and came into force in 2008 (NCCETC, 2018b) – it was available for systems below 100 kWh in capacity (DBEDT, 2019, 36).

NEM is designed for distributed generation systems where owners offset their electricity consumption from the grid with that of their PV system (Price et al., 2013; Eid et al., 2014; Watts et al., 2015; Ramírez et al., 2017). NEM works on the premise that customers are paid per kWh for the surplus electricity generated and sold to the grid at a rate commensurate with the prevailing retail tariff and utilities are legally obligated to buy this surplus energy (PUC, 2015; Watts et al., 2015; Lofthouse, Simmons & Yonk, 2015; EEI, 2016; Ramírez et al., 2017).

Households enrolled in programmes like NEM have their PV systems grid-connected to enable the technical flow of electricity to and from the house and wider electrical grid network. Such households are what Eid et al. (2014) and Georges et al. (2014) describe as 'prosumers', that is, agents that not only consume but also generate their electricity. Net consumers import more electricity from the grid than is generated from their distributed system and net producers export more than is consumed over the billing cycle. So, if a customer uses more electricity than is produced, that is, a net consumer, they only pay the utility for the net difference (DBEDT, 2015).

The NEM credit arrangement is dependent on the utility company and legal guidance facilitating the scheme (Hughes & Bell, 2006). This means that the crediting can use incumbent billing regimens and the disbursements can be made within this period or extend as a rolling credit on an annual basis (Dufo-López & Bernal-Agustín, 2015). In Hawai'i, excess credits are carried forward annually (PUC, 2015; DBEDT, 2015) and enables customers to discount their bills up to the point where consumption negates generation, that is, 'net-zero'.

Sources such as Cory, Couture and Kreycik (2009), Poullikkas, Kourtis and Hadjipaschaalis (2013) and Watts et al. (2015) further add that a conventional meter would only measure electricity consumption but NEM uses a bi-directional meter which also measures generation over the billing period. Practically, a conventional meter spins forward but NEM's bi-directional meter also spins backwards (measuring generation) and the difference here is the 'net energy' (Hughes & Bell, 2006; EEI, 2016).

With NEM, there is no direct feed-in to the grid (WEC, 2016). However, policies like FiTs use two meters (one for generation and consumption) and the energy exchanges are measured independently using different pricing designs for the imported and exported electricity; Hughes and Bell (2006), Mir-Artigues (2013) and Watts et al. (2015) state that the two meters are used because the exported and imported electricity are priced differently – which is in contrast to NEM.

The daily load curve for a 'typical' household, that is, the householders' energy services that create the demand on the energy system (Goss, 2010), is such there would be a small peak in the morning and larger one in the evening. Poullikkas, Kourtis and Hadjipaschaalis' (2013) narrative of daily energy use is an example that gives context to this generalization. They outline that the morning high occurs when the household awakes and begins their preparations for work and school by making breakfast and taking hot showers for instance. The midday trough occurs because no one is at home to use energy.

However, as the household returns from work and school in the evening, the demands are ramped up and peak at night when dinner is being prepared and television, computer and lighting use are high, for example. Air conditioning in tropical locations like Oʻahu may also contribute to the evening's peak – which Coffman et al. (2016, 8) show occurs between 5:00 pm and 9:00 pm in Hawaiʻi for instance. But there are two considerations for NEM worth touching on at this point. The first is perhaps the more obvious given the above generic insights and that is the disparity between peak solar energy generation at midday and the peak energy demand outlined.

Peak demand was detailed in Chapter 5 as the result of society's synchronized energy consumption patterns that create an increased period of consumption on a given day and which tend to occur around the same time each day. However, if PV under NEM is aimed at self-consumption and the household is not at home to use solar power when the generation is at its highest, this means that the power is sent to the grid. When the household returns home in the evening which is when solar power is not as productive, the demand for electricity outstrips the PV system's capacity to provide it so electricity is imported into the home from the grid.

Therefore, this technical disparity leads to considerations for load shifting, that is, moving peak demand load to off-peak periods (Sinha & De, 2016). An example of a common strategy used to achieve load shifting is time-of-use rates where different rates apply to different times of day and the highest rates apply during peak demand times (Sinha & De, 2016). This incentivizes householders to consume electricity during the day when solar power generation is at its highest and reduces the household's overall demand from the grid – and Hawaiʻi has applied such a residential pricing mechanism (Coffman et al., 2016).

Another is the use of batteries so that any solar power stored in the batteries during the day can be used during the peak demand periods which would, again, reduce the technical demands placed on the grid. Arik (2017) outlines that Hawaiʻi's drive to incentivize battery storage, as will be shown with the transition away from NEM to CSS, is part of a strategy to deal with the amounts of solar power coming into the grid relative to the disparity between peak energy demand and peak solar power generation.

The second consideration worth highlighting is more socio-economic and arguably not as obvious. With authors like Nolden (2012) acknowledging the increasing sensitization of government policies to social issues, the interviews with Oʻahu's policy officials flagged social inequity as one of the potential greatest weaknesses of the NEM programme. This inequity is based on the utility company's ratepayer distribution of NEM and non-NEM customers.

With PV under NEM, every unit of electricity saved is one that can be potentially sold back to the utility company (Goss, 2010). So, every kWh a solar homeowner generates is a kWh that the electric utility loses money from (Cunningham, 2014) at the crudest level. Additionally, beyond fuel costs, a utility company is a fixed-cost company so when large amounts of kWh are sold, there is a growth in sales and the fixed costs are spread out over a larger customer-base (Cohen, 2013) – and this reduces the per-unit cost across-the-board (Cohen, 2013).

However, NEM customers do not pay for the fixed costs when they generate their own power (Coffman et al., 2015; Wood, 2016). So, this shifts the burden to non-NEM customers and instead of being a tariff, NEM becomes a subsidy (Wood, 2016) because non-NEM customers subsidize the costs of using the grid for NEM customers (Cohen, 2013; Eid et al., 2014; Page, 2015). Coffman et al. (2015) further state that the utility's revenue decoupling in Oʻahu also enables this shift to occur.

An added dimension relates to utility companies' business models linked to NEM. Many NEM programmes have been perceived to decrease profits for utility companies which consequently lead them to adjust their electricity tariffs to compensate (Eid et al., 2014) – and several of the interviewees touched on this dilemma in Oʻahu as a point of contention. This notion means that since ratepayers are the financial base for a utility company, increasing retail tariffs would affect all those paying for the utility's services.

Nevertheless, the integration of PV under NEM may not decrease a utility's customer-count, but the proportional revenue contributions from within its ratepayer-base may potentially change such that NEM customers pay less given their bills' moderation through their PV. Cohen (2013) describes this inequity as the greatest implication of NEM's success. Further, lower-income families are the least likely to own residential solar (Couture et al., 2010) so this widens the inequity in the residential sector because higher-income households under NEM can benefit from the programme the most (potentially through increased electricity consumption, reduced electricity bills, socio-technical exclusivity and increased property values).

Despite this, NEM has been described as an economic, simple and rather straightforward means of integrating solar energy into a utility's business model (Poullikkas, Kourtis & Hadjipaschaalis, 2013). But from the householder's perspective, energy savings, personal budgets and payback periods are important considerations (Attari & Rajagopal, 2015) surrounding the classic trade-off between solar systems' capital costs and the returns from the energy generated (Bauner & Crago, 2015).

This trade-off is one of the reasons why solar adoption has been low in many islands, that is, there is uncertainty clouding whether the system will pay for itself within its lifetime (Bauner & Crago, 2015). The payback period and the internal rate of return are often used as proxies for determining the economic potential of solar (Dorf, 1984). Bauner and Crago outline that solar would be in good standing if conventionally sourced electricity prices are high and vice versa if it is low – as in Oʻahu and Trinidad respectively for instance. In Hawaiʻi the average payback for residential solar is six years (Peterson, 2017) and has an internal rate of return of 9% (Coffman et al., 2013).

Therefore, Hawaiʻi's high retail electricity rates combined with NEM made PV very attractive. This is especially so because NEM's one-to-one tariff rate encourages a high profitability for the owners of an enrolled solar system since the retail rate paid to the owner is higher than the rate that would have been received by a conventional generator for the same electricity (Watts et al., 2015). It also integrates well with other strategies like tax credits since this reduces the upfront costs of PV unlike NEM which is performance-based.

So, with due consideration for the eligible technology, customer tiers, maximum system size and utility business design, Stoutenborough and Beverlin (2008) and Cohen (2013) suggest that NEM is aimed at promoting the investment and expansion of distributed generation – and according to Gorak (2016), Hawaiʻi's NEM did its job of kick-starting renewable energy adoption.

However, the interviews conducted with policy officials in Oʻahu gave the impression that considerations such as the earlier-outlined social inequity debate mixed with arguments surrounding the solar market's maturity, circuit saturation and revenue loss for the utility company resulted in the presentation of a case before the PUC to review the NEM programme – and this consequently led to its closure in 2015.

Feed-in Tariffs

FiTs are one of the most popular renewable energy policies in the world (Cox & Esterly, 2016). They are a form of power procurement programme designed to increase the proportion of renewable energy in the energy mix (DeShazo & Matulka, 2009). DeShazo and Matulka (2009) further state that there is no set FiT design because it is adaptable, scalable and flexible; there is a global diversity of FiT designs which reflect a concomitant variety of policy goals (Couture et al., 2010).

FiTs legally obligate electric utilities to purchase renewable electricity from producers at an agreed rate through a long-term contract (Karekezi, Muzee & Corre, 2009; Cory, Couture & Kreycik, 2009; Cornfeld & Sauer, 2010; Couture et al., 2010; Cox & Esterly, 2016). The contract is usually between 10 to 25 years (Mostafa, 2014; Pyrgou, Kylili & Fokaides, 2016) and Hawaiʻi's is 20 (Coffman et al., 2015, 11).

Sources such as Cory (2009), Couture et al. (2010), Cornfeld and Sauer (2010), and Cox and Esterly (2016) state that FiTs should detail elements such as eligible technologies, technology generation capacity, project size, location and resource quality. Cox and Esterly (2016) also state that clearly specifying the participating actors, their roles and the respective procedures guiding their interactions can save time and finances. This is important because there is an initial administrative burden involved in setting up FiTs (Cory, Couture & Kreycik, 2009).

However, Karekezi, Muzee and Corre (2009), Couture et al. (2010), Cornfeld and Sauer (2010), and Cox and Esterly's (2016) work describe FiTs' basic components: a payment per kWh of generated and exported electricity, the long-term contract and priority access to the grid for the renewable energy generated. The last component ensures that the energy is consumed before other sources.

Hawai'i's FiT was introduced in 2009 and its rates were approved by the PUC in 2010/2011 (EEEI, 2014, 76). It is designed for generators smaller than 5 MW (DBEDT, 2019, 39) and runs as three separate programmes under the HEI (NCCETC, 2018a) that amount to an aggregated 80 MW (EEEI, 2014, 76; DBEDT, 2019, 39) (of which 75% is in O'ahu). It is worth recalling that earlier in the chapter, the HEI was described as made up of the parent HECO company and its subsidiaries HELCO and MECO. Therefore, the different islands under each company's remit would be under different FiT jurisdictions.

But with the pros and cons of FiTs usually being economic (Couture et al., 2010), the real detail lies within the payment design. The FiT rate is usually set at a level that provides a reasonable rate of return over the contract (Cornfeld & Sauer, 2010) and high enough to attract investors (Mostafa, 2014) – though longer contracts may occasionally have lower rates (Cox & Esterly, 2016).

Hughes and Bell (2006) outline that grid-connected policies can allow the utility to buy back a household's solar power free-of-charge, below the retail rate, at the retail rate or at a premium which is higher than the retail rate and usually offered for specific technologies. In that regard, setting the FiT rate is perhaps the most pivotal and difficult element because setting a 'reasonable' rate seeks to balance-out cost recovery and customer profit (Couture et al., 2010; Nolden, 2012; Cox & Esterly, 2016). This is why Cory, Couture and Kreycik (2009) state that detailed analyses are required from the onset to set the right payment level(s).

In Hawai'i, the rate for PV under Tier 1 of the FiT in all the HEI's islands was set at either US21.8 cents or 27.4 cents per kWh; under Tier 2, either 18.9 cents or 23.8 cents in O'ahu, Maui and Hawai'i Island, and L'na'i and Moloka'i; and under Tier 3, either 19.7 cents or 23.6 cents in O'ahu, and Maui and Hawai'i Island (DBEDT, 2019, 39). These tiers and rates are associated with the size of the PV systems (Coffman et al., 2015) and specifically for the case study, O'ahu, the limit to Tier 1 is 20 kW; 500 kW for Tier 2;

and 5 MW for Tier 3 (EEEI, 2014, 77; DBEDT, 2019, 39). Most households will therefore be under the Tier 1 rating (Coffman et al., 2015).

A question that may be raised at this point is why there two different rates under the tiers. The EEEI (2014) and the DBEDT (2019) outline that in Hawaiʻi, customers were able to qualify for either a 24.5% or 35% State tax rebate/credit so where a customer accepts the lower of the two, then they are eligible for the higher FiT rate and vice versa.

This policy integration represents a small trade-off between capital cost reduction and performance-based incentivization. But it shows the agency of the government and utility company integrating in order to present a flexible policy solution. This also means that as would be the case under NEM, FiT customers must stand the PV system's upfront costs (Cory, 2009; Cory, Couture & Kreycik, 2009; Couture et al., 2010; Blechinger & Shah, 2011). So once again, tax incentives can be useful instruments here.

Cory, Couture & Kreycik (2009), Couture et al. (2010), Kreycik (2011) and Cox and Esterly (2016) outline that FiTs' rate-determinations can be based on levelized costs, avoided costs, resource quality or even auction-based considerations. Levelized costs are usually based on factors such as capital costs, installation, operations and maintenance and decommissioning (Chivers, 2015). Avoided cost considerations are based on the premise that since the utility does not stand the costs of certain developments, for example the health impacts from emissions, climate change or installing household PV, this can be economically valuated to help determine FiT rates. With geographically variable resources, these variations can be marked enough to make resource quality a rate-setting consideration. Auction-based determinations are done via tendering processes.

FiT work such as Couture et al. (2010), Cornfeld and Sauer (2010), Pyrgou, Kylili and Fokaides (2016), and Cox and Esterly (2016) further outline that FiT payments can be either fixed or premium cost designs; the fundamental distinction is whether the payments are linked to the electricity market price in which case the fixed version is independent of a market's prices.

Nevertheless, consideration should be extended towards how FiTs evolve over time because they can be modified to reflect changing market conditions and these changes can be pre-determined or they can follow comprehensive reviews and subsequent adjustments which can be based on achieving milestones or following regular revision regimens (DeShazo & Matulka, 2009).

Cory (2009) states that degressions, that is, decreasing FiT rates over time, can be used in response to the industry's potential to become complacent and dependent on FiTs for developing projects (what she referred to as a policy crutch); tariff degressions usually stimulate market innovations and incentivizes companies' technology cost-reductions (Couture et al., 2010). However, if the degression is poorly designed then the market and industry can be severely affected.

FiT prices can also change over time to adapt to changes in generation costs relative to the cost of the technology (Cory, Couture & Kreycik, 2009;

EEEI, 2014; Cox & Esterly, 2016) or even inflation (Couture et al., 2010). The EEEI (2014) states, however, that once a customer enters into a contract with their utility, the agreed-to rate is guaranteed over the stipulated period, irrespective of changes that occur within this period. Therefore, the modifications refer to the FiT programme versus an individual's FiT contract.

If the FiT's rate is reduced after a revision then it represents a responsive tariff degression (Couture et al., 2010). Couture et al. (2010) and Cox and Esterly (2016) also outline that policymakers may alternatively establish a pre-determined tariff degression. Further, Mostafa (2014) shows that degressions can be implemented along a constant annual decline profile or even done in a step-like fashion over the lifetime of the FiT.

Nevertheless, a FiT's revision regimen should reflect evolving conditions (Cory, Couture & Kreycik, 2009), and any changes should be transparent and predictable (Cox & Esterly, 2016). But revising a FiT and adjusting its rates will create uncertainty (WEC, 2016) so any revisions need to take this into account because the solar energy industry needs certainty to thrive and customers need guarantees to build confidence in their investments. Additionally, FiTs' financing should be stable and guaranteed so that there is customer-confidence surrounding the payment plans' longevity (Cox & Esterly, 2016).

When Hawai'i's FiT was implemented, it was done so as a two-year pilot (Morita, Akiba & Champley, 2014, 1). So, given the above thoughts on FiT revisions, Hawai'i's FiT has been closed to new applications since 2014 (DBEDT, 2019; HECO, 2019d). The end of FiT followed by NEM (in 2015) therefore suggests that the State underwent, and arguably still is undergoing, major changes in its energy policy environment which reflect the changing conditions that the regime described earlier in the chapter is facing.

Transitioning from photovoltaics to batteries

Despite the closure of policies such as NEM (and the FiT) in Hawai'i, there are several policies that maintain grid-connected PV's place in the wider energy transitions narrative, for example standard interconnection agreements which is in addition to those PV systems still under contract through FiT and NEM. However, what is worth pointing out is that notable post-NEM policy initiatives are hinting at O'ahu's next energy transition: the CGS and CGS Plus; CSS; NEM Plus; and Smart Export schemes are the main policies here.

Simply put, CGS and CSS are a fixed-rate, grid-connected policy and self-consumption incentive scheme respectively (Gorak, 2016; Arik, 2017; Paidipati & Romano, 2017). Paidipati and Romano (2017), the HECO (2017a, b) and the HECO (2019e) point out that the main differences between NEM and CGS are: CGS' rate is lower and fixed unlike NEM which was set at the retail rate and fluctuated with it, and there is no annual credit rollover as was the case with NEM. In the case of CSS, it allows customers to install rooftop PV that do not export power to the grid, so all the power must be consumed directly or stored for later use (HECO, 2019f).

NEM Plus enables existing NEM customers to increase their installed capacity but with the caveat that the added capacity is geared towards self-consumption and existing capacity retains the agreement under the original NEM scheme (DBEDT, 2019). The DBEDT (2019) also outlines that Smart Export is more specifically for battery-integrated household PV systems where the batteries are to be charged by the PV system during the day and discharged at night. It also enables customers to export power at night if there is a surplus and to be credited for this export.

So, thanks to policies like these, the DBEDT (2019, 36) states that the distributed energy technologies (most of which are likely to be PV) enrolled in them make up the largest proportion of the State's renewable energy at 39.6% as well as 10.9% of all electricity sales. In that same train of thought, it was also reported that of the total 74,331 residential rooftop installations statewide, Oʻahu has at least 51,087 versus 11,414 in Maui and 11,830 on Hawaiʻi Island (DBEDT, 2019, 36).

However, the last three policy options highlighted, that is, CSS, NEM Plus and Smart Export show that battery storage is becoming a more prominent technological and policy-based feature in Oʻahu (and Hawaiʻi). Therefore, the policy transitions away from the FiT and NEM to the latter ones outlined are linked to a technological transition from grid-connected household PV to/through battery-integrated household PV.

Summary

Oʻahu is another example of a small, tropical island which relies on significant amounts of imported oil/petroleum to satisfy its energy needs – which includes power generation. Like Barbados, household solar water heating was around since the 1970s. However, unlike Barbados, which is now starting its PV transition, Oʻahu has already transitioned to/through distributed residential PV (during the mid-2000s).

Such household PV systems work outside the conventional technical electricity system as a disruptive stimulus. A conventional, centralized power generation regime (like Trinidad's) that integrates distributed PV becomes more complex. Households are important actors that will be affected by this increased complexity because incentivizing distributed PV through solar energy policies increases the interactivity between households and utility company for instance.

The conventional design involves households consuming electricity and paying the utility for this electricity based on the bills issued. With the introduction of grid-connected incentives, households can be paid for the solar power they generate and export, and the utility company gains added power on their grid – which is in addition to the original interactions between both actors. Distributed household PV therefore introduce new dimensions of electricity generation, transmission and distribution; financing and expenditures; fixed-cost recovery; and customer-service.

The integration of such residential solar energy constitutes an industrial transition as the conventional energy industry made up of hydrocarbon and power generation companies transition into an industry that includes solar energy retailers and households. There is also a technological transition tied to this where large-scale centralized power generation units transition into a system that includes decentralized household units.

Many factors enabled the mainstreaming of PV in Oʻahu and they were not all 'on-island' forces. They include international oil price highs, global economic recessions and cheaper international solar panel manufacturing costs, for instance. At the local level, the high retail electricity rates, improvements in energy efficiency and the implementation of policies helped mainstream solar energy. Relative to the last aspect, policies, tax incentives, RPSs (and EEPSs), the FiT and NEM were and are amongst the most influential policies implemented.

Solar energy tax credits have been around in Hawaiʻi since solar water heating took off and it has expanded to include eligible PV systems over the lifetime of the policy – which is in addition to the historical variation in the tax credit allowances and additional allowances from the Federal Government being added to those from the State. Such variations are fluctuant cycles that mark a form of policy oscillation during the (solar) energy transition. Nevertheless, the policy reduced/reduces the high capital costs of solar systems for households.

RPSs are broad, legislated renewable energy ambitions that the State has committed to achieving; they are currently set at achieving 100% renewable energy generation/sales by 2045 and use previous RPS goals as interim targets, for example '40% by 2040'. Increasing these targets requires inputs that may or may not lie within the regime and thus modifies it to facilitate the actors and resources needed to provide this added input. Increasing targets in this way also represents an intra-policy transition based on quantitative modifications. The RPSs are also complemented by EEPSs which support energy efficiency improvements.

The FiT was introduced as a pilot and alternative to NEM. It ran as three programmes under the HEI. Customers enrolled in the FiT have 20-year power purchase contracts for the solar power they generate and export to the grid. The FiT has three tiers and household PV fits within the first tier, that is, systems less than 20 kW in capacity. In the wider scope, FiTs traditionally use a generation and consumption meter because the outgoing and incoming electricity is priced differently. The FiT rate can be based on elements such as resource quality, avoided costs, auction process or even levelized costs. They can also be fixed or premium cost designs where the fundamental difference is whether the rates are linked to the electricity market prices; the latter are dependent on the market's prices.

Unlike a FiT, NEM uses a bi-directional meter and offers the owner of an eligible PV system a credit per kWh for surplus electricity sent into the grid at a rate that is equivalent to their retail electricity rate from the utility company. Further, the NEM in Hawai'i uses an annual rolling credit where the credit-debit determinations are made at the end of the year based on the solar power generated and electricity consumed. NEM is also designed to enable households to consume their solar power so the electricity consumed by the household will be a combination of both solar power and the conventional supply from the grid. So, relative to the sale of surplus electricity in Hawai'i, households are also only able to gain credits for the excess solar power that amounts to their consumption from the grid, that is, net-zero.

However, with there being considerations and debates surrounding market maturity, technical circuit saturation, peak energy demand and social inequity, the State closed the FiT and NEM and transitioned to options that include CGS, CGS Plus, NEM Plus, CSS and the Smart Energy scheme. Such developments are examples of inter-policy transitions. Further, several are geared towards encouraging solar self-consumption and that has opened the way for batteries to become more widely integrated parts of O'ahu's residential energy system. Therefore, O'ahu's regime is now beginning its transition to/through battery-integrated household PV.

Grid-connected policies are without a doubt one of the most popular strategies for mainstreaming solar energy. But no matter the design, defining the eligible technologies, the technologies' details and the facilitating actors together with their roles and responsibilities are key areas that need to be well thought out and transparent. Pilots are also valuable ways of introducing policies. It goes without saying that determining equitable and attractive rates under a grid-connected policy scheme is perhaps the most pivotal element of its success, however.

Governments have a major role through policy implementation, but policies' success depend on the wider institutional network because they are not standalone features of an energy transition. This comes in light of such energy transitions involving new and more relationships between households, installers and industry, as well as policy-influencers. In the case of the last, this particularly applies to institutions such as the HSL, PUC and HSEO in Hawai'i.

Institutions' influences on and in their regime's electricity networks are related to their realized agency. So, with changes in the way in which actors in the regime interact, one can expect there to be changes in the power distribution and functionality of actors as well. This is significant to consider since it gives insight into how power and agency distributions can thwart or support solar in small, tropical islands – and working within the present regime's 'power' structure may be the path of least resistance.

This chapter marks the end of an empirical journey that began in the Caribbean with Trinidad then Barbados and has now ended in the Pacific on O'ahu. Of the three islands, O'ahu is the furthest along in the residential solar energy transition and so its experience with household PV and policymaking offers good insights for other islands wrestling with their own PV transition. But with the narrative presented from Chapter 4 through to this one, are some conclusions on mainstreaming solar energy that should be presented, and Chapter 8 presents them.

References

ABB. (2012). *Integrating Renewables into Remote or Isolated Power Networks and Micro Grids: Innovative Solutions to Ensure Power Quality and Grid Stability*. Retrieved from https://library.e.abb.com/public/3a9c2c4eac30411d8a965e9bde014382/3BUS095580_Microgrids%20Overview%20Brochure_141223.pdf.

Alm, R. (2015). *Hawaii's 100% Renewable Goal: The Confluence of Policy and Reality*. Paper presented at the International Conference on Perspectives on the Development of Energy and Mineral Resources Hawai'i, Mongolia and Germany, Honolulu, Hawaii. Retrieved from http://socialsciences.hawaii.edu/conference/demr2015/_papers/alm-robert.pdf.

Arik, A. D. (2017). *Residential Battery Systems and The Best Time to Invest. A Case Study of Hawaii*. Honolulu, State of Hawaii: The Economic Research Organization at the University of Hawai'i. Retrieved from http://uhero.hawaii.edu/assets/WP_2017-9.pdf.

Attari, S. Z. & Rajagopal, D. (2015). Enabling Energy Conservation Through Effective Decision Aids. *Journal of Sustainability Education*, 8. Retrieved from www.jsedimensions.org/wordpress/content/enabling-energy-conservation-through-effective-decision-aids_2015_01/.

Baker, B. (2014, February 19). How Hawaii is Transitioning from Oil Dependence to Solar Energy. *EcoWatch*. Retrieved from http://ecowatch.com/2014/02/19/hawaii-oil-solar/.

Bass, S. & Dalal-Clayton, B. (1995). *Small Island States and Sustainable Development: Strategic Issues and Experience*. Environmental Planning Issues, 8, Environmental Planning Group, International Institute for Environment and Development, London. Retrieved from http://pubs.iied.org/pdfs/7755IIED.pdf.

Bauner, C. & Crago, C. (2015). Adoption of Residential Solar Power Under Uncertainty: Implications for Renewable Energy Incentives. *Energy Policy*, 86: 27–35. Retrieved from https://doi.org/10.1016/j.enpol.2015.06.009.

Beerepoot, M. & Beerepoot, N. (2007). Government Regulation as an Impetus for Innovation: Evidence from Energy Performance Regulation in the Dutch Residential Building Sector. *Energy Policy*, 35(10): 4812–4825. Retrieved from https://doi.org/10.1016/j.enpol.2007.04.015.

Blechinger, P. F. H. & Shah, K. U. (2011). A Multi-criteria Evaluation of Policy Instruments for Climate Change Mitigation in the Power Generation Sector of Trinidad and Tobago. *Energy Policy*, 39(10): 6331–6343. Retrieved from https://doi.org/10.1016/j.enpol.2011.07.034.

Braccio, R. & Finch, P. (2011). *HCEI Roadmap 2011 Edition*. National Renewable Energy Laboratory, United States Department of Energy. Retrieved from www.nrel.gov/docs/fy11osti/52611.pdf.

Brooks, B. (2003). *Installing and Inspective Solar Photovoltaics (PV) System for Code Compliance*. Presentation given at a Workshop sponsored by the Hawaii State Energy Office, Hawaiian Electric Company and the City and County of Honolulu, Oahu,

Hawaii. Retrieved from http://energy.hawaii.gov/wp-content/uploads/2011/10/Installing-Inspecting-PV-Systems_2003.pdf.

Burnett, H. S. (2011). *Solar Power Prospects*. Policy Report No. 334. Dallas, TX: National Center for Policy Analysis. Retrieved from www.ncpathinktank.org/pub/st334.

Chabrol, M. (2016). Re-examining Historical Energy Transitions and Urban Systems in Europe. *Energy Research and Social Science*, *13*: 194–201. Retrieved from https://doi.org/10.1016/j.erss.2015.12.017.

Chivers, D. (2015). *Renewable Energy: Cleaner, Fairer Ways to Power the Planet*. Oxford, UK: New International Publications Limited.

Codiga, D. A. (2013). Hot Topics in Hawaii Solar Energy Law. *Hawaii Bar Journal*, *17*(5): 4–15. Retrieved from www.honolulu-lawyers.com/docs/dac_article_may_2013_hsba.pdf.

Coffman, M., Allen, S. & Wee, S. (2018). *Determinant of Residential Solar Photovoltaic Adoption*. Honolulu, State of Hawaii: The Economic Research Organization at the University of Hawaii. Retrieved from www.uhero.hawaii.edu/assets/WP_2018-1.pdf.

Coffman, M., Bernstein, P., Wee, S. & Arik, A. (2016). *Estimating the Opportunity for Load-Shifting in Hawaii: An Analysis of Proposed Residential Time-of-Use Rates*. Honolulu, State of Hawaii: The University of Hawaii Economic Research Organization at the University of Hawaii. Retrieved from www.uhero.hawaii.edu/assets/TOU-Rates_8-2.pdf.

Coffman, M., Bonham, C., Wee, S. & Salim, G. (2013). *Tax Credit Incentives for Residential Solar Photovoltaics in Hawai'i*. Honolulu, State of Hawaii: The Economic Research Organization at the University of Hawaii. Retrieved from www.uhero.hawaii.edu/products/view/409.

Coffman, M., Fripp, M., Roberts, M. J. & Tarui, N. (2015). *Efficient Design of Net Metering Agreements in Hawaii and Beyond*. Honolulu, State of Hawaii: The Economic Research Organization at the University of Hawaii. Retrieved from https://uhero.hawaii.edu/assets/Net_Metering.pdf.

Cohen, R. (2013). The Giant Headache that is Net Energy Metering. *The Electricity Journal*, *26*(6): 1, 5–7. Retrieved from https://doi.org/10.1016/j.tej.2013.06.010.

Cole-Brooks, L. (2014). *Hawaii Solar Energy Association*. Presentation published by the Chamber of Commerce Hawaii. Retrieved from www.cochawaii.org/wp-content/uploads/HSEA-Honolulu-Chamber-of-Commerce.pdf.

Corbus, D., Kuss, M., Piwko, D., Hinkle, G., Matsuura, M., McNeff, M., Roose L. & Brooks, A. (2013). All Options on the Table: Energy Systems Integration on the Island of Maui. *Power and Energy Magazine, IEEE*, *11*(5): 65–74. Retrieved from https://doi.org/10.1109/MPE.2013.2268814.

Cornfeld, J. & Sauer, A. (2010). *Feed-in Tariffs*. Issue Brief. Environmental and Energy Study Institute. Retrieved from www.eesi.org/files/feedintariff_033110.pdf.

Cory, K. (2009). *Renewable Energy Feed-in-Tariffs: Lessons Learnt from the U.S. and Abroad*. Presentation published by the National Renewable Energy Laboratory. Retrieved from http://ledsgp.org/resource/renewable-energy-feed-in-tariffs-lessons-learned-from-the-us-and-abroad-presentation/?loclang=en_gb.

Cory, K., Couture, T. & Kreycik, C. (2009). *Feed-in Tariff Policy: Design, Implementation, and RPS Policy Interactions*. Colorado: National Renewable Laboratory. Retrieved from www.nrel.gov/docs/fy09osti/45549.pdf.

Cory, K. S. & Swezey, B. G. (2007). *Renewable Portfolio Standards in the States: Balancing Goals and Implementation Strategies*. Colorado: National Renewable Energy Laboratory. Retrieved from www.nrel.gov/docs/fy08osti/41409.pdf.

Couture, T. D., Cory, K., Kreycik, C. & Williams, E. (2010). *Feed-in-Tariffs vs Feed-in-Premium Policies*. In *A Policymaker's Guide to Feed-in* Tariff Policy Design. Colorado: National Renewable Energy Laboratory. Retrieved from http://helapco.gr/pdf/FiT_vs_FiP_NREL.pdf.

Cox, S. & Esterly, S. (2016). *Feed-in Tariffs: Good Practises and Design Considerations. A Clean Energy Regulators Initiative Report*. Colorado: National Renewable Energy Laboratory. Retrieved from www.nrel.gov/docs/fy16osti/65503.pdf.

Critz, D. K., Busche, S. & Connors, S. (2013). Power Systems Balancing with High Penetration Renewables: The Potential of Demand Response in Hawaii. *Energy Conversion and Management*, 76: 609–619. Retrieved from https://doi.org/10.1016/j.enconman.2013.07.056.

Cunningham, N. (2014). (November 24). This Company Could Revolutionize Energy Utilities. *Oilprice.com*. Retrieved from http://oilprice.com/Energy/Energy-General/This-Company-Could-Revolutionize-Energy-Utilities.html.

DBEDT (Department of Business, Economic Development and Tourism). (2019). *Hawaii Energy Facts & Figures July 2019*. Honolulu, State of Hawaii: Hawaii State Energy Office. Retrieved from https://energy.hawaii.gov/wp-content/uploads/2019/07/2019-FF_Final.pdf.

DBEDT (Department of Business, Economic Development and Tourism). (2014). *Hawaii Energy Facts and Figures November 2014*. Honolulu, State of Hawaii: Hawaii State Energy Office. Retrieved from http://energy.hawaii.gov/wp-content/uploads/2014/11/HSEO_FF_Nov2014.pdf.

DBEDT (Department of Business, Economic Development and Tourism). (2015). *Hawaii Energy Facts and Figures May 2015*. Honolulu, State of Hawaii: Hawaii State Energy Office. Retrieved from http://energy.hawaii.gov/wp-content/uploads/2011/10/HSEO_FF_May2015.pdf.

DBEDT (Department of Business, Economic Development and Tourism). (2017). *Hawaii Energy Facts & Figures May 2017*. Honolulu, State of Hawaii: Hawaii State Energy Office. Retrieved from https://energy.hawaii.gov/wp-content/uploads/2011/10/HSEOFactsFigures_May2017_2.pdf.

DBEDT (Department of Business, Economic Development and Tourism). (2003). *Have Some Energy on The House … Solar: Questions and Answers on Solar Water Heating*. State of Hawaii, Honolulu. Retrieved from http://energy.hawaii.gov/wp-content/uploads/2011/10/HaveEnergyOnHouse2003.pdf.

DeShazo, J. R. & Matulka, R. (2009). *Best Practices for Implementing a Feed-in Tariff Program*. Los Angeles, CA: Luskin Center for Innovation, Luskin School of Public Affairs, University of California. Retrieved from https://luskin.ucla.edu/sites/default/files/Best%20Practices%20for%20Implementing%20a%20Feed%20in%20Tariff%20Program.pdf.

Dorf, R. C. (1984). Managerial and Economic Barriers and Incentives to the Commercialization of Sola Energy Technologies. *Engineering Management International*, *2*(1): 17–31. Retrieved from https://doi.org/10.1016/0167-5419 (84)90034-6.

Dufo-López, R. & Bernal-Agustín, J. L. (2015). A Comparative Assessment of Net Metering and Net Billing Policies. Study Cases for Spain. *Energy*, *84*: 684–694. Retrieved from https://doi.org/10.1016/j.energy.2015.03.031.

EEEI (Energy and Environmental Economics Incorporated). (2014). *Evaluation of Hawaii's Renewable Energy Policy and Procurement Final Report January 2014 Revision*. San Francisco, CA. Retrieved from http://puc.hawaii.gov/wp-content/uploads/2013/04/HIPUC-Final-Report-January-2014-Revision.pdf.

EEI (Edison Electric Institute). (2016). *Solar Energy and Net Metering*. Washington, DC. Retrieved from www.eei.org/issuesandpolicy/generation/NetMetering/Documents/Straight%20Talk%20About%20Net%20Metering.pdf.

Eid, C., Guillén, J. R., Marín, P. F. & Hakvoort, R. (2014). The Economic Effect of Electricity Net-metering with Solar PV: Consequences for Network Cost Recovery, Cross Subsidies and Policy Objectives. *Energy Policy*, *75*: 244–254. Retrieved from https://doi.org/10.1016/j.enpol.2014.09.011.

Electricity Local. (2019). *Retail Electricity Rates & Consumption in Hawaii*. Retrieved from www.electricitylocal.com/states/hawaii/.

GEC (Global Energy Concepts LLC). (2006). *A Catalog of Potential Sites for Renewable Energy in Hawaii*. Honolulu: State of Hawaii Department of Land and Natural Resources, and Department of Business, Economic Development and Tourism. Retrieved from http://energy.hawaii.gov/wp-content/uploads/2011/10/A-Catalog-of-Potential-Sites-for-Renewable-Energy-in-Hawaii.pdf.

General Electric (GE) Energy Consulting. (2015). *Technical Report: Hawaii Renewable Portfolio Standards Study*. Hawaii Natural Energy Institute. Retrieved from www.hnei.hawaii.edu/sites/www.hnei.hawaii.edu/files/Hawaii%20RPS%20Study%20-%20Final%20Report.pdf.

Georges, E., Braun, J. E., Groll, E., Horton, W. T. & Lemort, V. (2014). *Impact of Net Metering Programs on Optimal Load Management in US Residential Housing – A Case Study*. Paper presented at the 9th International Conference on System Simulation in Buildings, Liège, Belgium. Retrieved from http://orbi.ulg.ac.be/bitstream/2268/178419/1/P64v2.pdf.

Gorak, T. C. (2016). *Advancing Renewables: Lessons Learned in Hawaii (So Far)*. Presentation given at the 2016 Energy Information Administration Conference, Washington, DC. Retrieved from www.eia.gov/conference/2016/pdf/presentations/gorak.pdf.

Goss, B. (2010). *Choosing Solar Electricity: A Guide to Photovoltaic Systems*. Machynlleth, UK: Centre for Alternative Technology.

Haleakala Solar. (2014, January 27). *Brief History of Solar and Hawai'i Photovoltaic Highlights*. Retrieved from www.haleakalasolar.com/Hawai'i-solar/brief-history-of-solar-Hawai'i-photovoltaic-highlights/.

HECO (Hawaii Electric Company Incorporated). (2011). *Fuel Oil Use in Hawaii*. Retrieved from www.hawaiianelectric.com/Documents/about_us/company_facts/fuel_oil_use_042011.pdf.

HECO (Hawaiian Electric Company Incorporated). (2017a). *Customer Grid-Supply (CGS) Billing.* Retrieved from www.hawaiianelectric.com/documents/products_and_services/customer_renewable_programs/CGS_credits_2017.pdf.
HECO (Hawaiian Electric Company Incorporated). (2017b). *Hawaiian Electric Companies Open Up Capacity for Grid-supply Solar Program.* Retrieved from www.hawaiielectriclight.com/hawaiian-electric-companies-open-up-capacity-for-grid-supply-solar-program.
HECO (Hawaiian Electric Company Incorporated). (2019a). *Solar Energy.* Renewable Energy Sources. Retrieved from www.hawaiianelectric.com/clean-energy-hawaii/our-clean-energy-portfolio/renewable-energy-sources/solar.
HECO (Hawaiian Electric Company Incorporated). (2019b). *Power facts.* Retrieved from www.hawaiianelectrc.com/about-us/power-facts.
HECO (Hawaiian Electric Company Incorporated). (2019c). *Quarterly Installed Solar Data: 3rd Quarter, 2019.* Retrieved from www.hawaiianelectric.com/clean-energy-hawaii/our-clean-energy-portfolio/quarterly-installed-solar-data.
HECO (Hawaiian Electric Company Incorporated). (2019d). *Feed-in Tariff Program.* Selling Power to the Utility. Retrieved from www.hawaiianelectric.com/clean-energy-hawaii/selling-power-to-the-utility/feed-in-tariff.
HECO (Hawaiian Electric Company Incorporated). (2019e). *Customer Grid-Supply.* Customer Renewable Programs. Retrieved from www.hawaiianelectric.com/clean-energy-hawaii/producing-clean-energy/customer-renewable-programs/customer-grid-supply.
HECO (Hawaiian Electric Company Incorporated). (2019f). *Customer Self-Supply.* Customer Renewable Programs. Retrieved from www.hawaiianelectric.com/clean-energy-hawaii/producing-clean-energy/customer-renewable-programs/customer-self-supply.
HNEI (Hawaii Natural Energy Institute). (2008). *Assessment of the State of Hawaii's Ability to Achieve 2010 Renewable Portfolio Standards- Final Report.* Honolulu, State of Hawaii: School of Oceanography and Earth Science and Technology, University of Hawaii. Retrieved from http://puc.hawaii.gov/wp-content/uploads/2013/04/RTL-2010RenewablePortStd.pdf.
HNEI (Hawaii Natural Energy Institute). (2016). *World Bank Study Tour.* Presentation published by the School of Ocean and Earth Science and Technology, University of Hawaii, Honolulu, State of Hawaii. Retrieved from www.esmap.org/sites/esmap.org/files/HNEI%20-%20World%20Bank%20Study%20Tour%2005242016_web.pdf.
Hollier, D. (2015, February 10). Which NextEra Will Hawaii Get? *Hawaii Business.* Retrieved from www.hawaiibusiness.com/which-nextera-will-hawaii-get/.
HSEA (Hawaii Solar Energy Association). (2012). *About HSEA.* Retrieved from www.hsea.org/about.
Hughes, L. & Bell, J. (2006). Compensating Customer-generators: A Taxonomy Describing Methods of Compensating Customer-generators for Electricity Supplied to the Grid. *Energy Policy, 34*(13): 1532–1539. Retrieved from https://doi.org/10.1016/j.enpol.2004.11.002.
Hunter, L., Westlake, D. & Griffith, J. (2013). Renewable Energy in Hawaii. In *Renewable Energy in the 50 States: Western Region, 2013 Edition* (pp. 14–15). American Council on Renewable Energy, Washington, DC. Retrieved from

https://energy.hawaii.gov/wp-content/uploads/2011/09/2013-RE-50-States_ACORE.pdf.
IEC (International Electrotechnical Commission). (2012). *Grid Integration of Large-capacity Renewable Energy Sources and Use of Large-capacity Electrical Energy Storage.* Geneva, Switzerland. Retrieved from www.iec.ch/whitepaper/pdf/iecWP-gridintegration largecapacity-LR-en.pdf.
IER (Institute for Energy Research). (2013). *Hawaii: An Energy and Economic Analysis.* Retrieved from http://instituteforenergyresearch.org/wp-content/uploads/2013/08/State-Analysis-Hawaii.pdf.
InSynergy Engineering. (2012). *State of Hawaii Solar Water Heating Impact Assessment (1992–2011).* Honolulu, State of Hawaii: Hawaii Department of Business, Economic Development and Tourism. Retrieved from http://energy.hawaii.gov/wp-content/uploads/2011/10/STATE-OF-HAWAII-SOLAR-WATER-HEATING-IMPACT-ASSESSMENT-_1992-2011.pdf.
Jensen, T. L. (2000). *Renewable Energy on Small Islands* (2nd ed.). Forum for Energy and Development. Retrieved from www.gdrc.org/oceans/Small-Islands-II.pdf.
Johnson, J. & Chertow, M. (2009). *A Systems Approach to Energy Sustainability in Hawai'i County.* Paper presented at the 47th Hawaii International Conference on System Sciences, Waikiloa, Big Island, Hawaii. Retrieved from https://doi.ieeecomputersociety.org/10.1109/HICSS.2009.553.
Karekezi, S., Muzee, K. & Corre, J. (2009). *The Role of Feed-in Tariff Policy in Renewable Energy Development in Developing Countries: A Toolkit for Parliamentarians.* Nairobi, Kenya: Energy, Environment and Development Network for Africa. Retrieved from https://agora-parl.org/node/8953.
Kaya, A. & Yalcintas, M. (2010). Energy Consumption Trends in Hawaii. *Energy, 35*(3): 1363–1367. Retrieved from https://doi.org/10.1016/j.energy.2009.11.019.
Kreycik, C., Couture, T. B. & Cory, K. (2011). *Innovative Feed-in Tariff Designs that Limit Policy Costs.* Colorado: National Renewable Energy Laboratory. Retrieved from www.nrel.gov/docs/fy11osti/50225.pdf.
Lim, R. (2011). *DEBT Energy Plan.* Presentation given at the Hawaii Clean Energy Initiative Plenary Session, Honolulu, State of Hawaii. Retrieved from www.hawaiicleanenergyinitiative.org/storage/media/1-Lim.pdf.
Lofthouse, J., Simmons, R. T. & Yonk, R. M. (2015). *Reliability of Renewable Energy: Solar.* Utah: Institute of Political Economy, Utah State University. Retrieved from www.usu.edu/ipe/wp-content/uploads/2015/11/Reliability-Solar-Full-Report.pdf.
Lyon, T. P. & Yin, H. (2010). Why Do States Adopt Renewable Portfolio Standards? An Empirical Investigation. *The Energy Journal, 31*(3): 133–157. Retrieved from www.jstor.org/stable/41323297.
Marumoto, C. (2013). *Guide to Government in Hawaii (14th ed.).* Legislative Reference Bureau, State Capitol, Honolulu, State of Hawaii. Retrieved from http://lrbhawaii.org/gd/gdgovhi.pdf.
Mir-Artigues, P. (2012). The Spanish Regulation of the Photovoltaic Demand-side Generation. *Energy Policy, 63*: 664–673. Retrieved from https://doi.org/10.1016/j.enpol.2013.09.019.
Morita, H. M. (2014). *State Examples of Innovation & Stakeholder Engagement. Hawaii Public Utilities Commission. Utility Business Models that Align with State Clean Energy Goals.* Pre-

sentation given at the National Governors Association Workshop, Washington, DC. Retrieved from http://nga.org/files/live/sites/NGA/files/pdf/2014/1409UtilityBusinessModelsStateCleanEnergyControl_Morita.pdf.

Morita, H., Akiba, L. H. & Champley, M. E. (2014). *PUC Approves Joint Plan on the Administration of FIT Program Queues to Accelerate Project Completions*. Honolulu, State of Hawaii: Public Utilities Commission. Retrieved from www.hawaiianelectric.com/documents/clean_energy_hawaii/producing_clean_energy/fit/public_utilities_commission_press_release_dated_december_8_2014.pdf.

Morita, M. & Mangelsdorf, M. (2015, July 6). It's Time to End Net Energy Metering in Hawaii. *Greentech Media*. Retrieved from www.greentechmedia.com/articles/read/Its-Time-to-End-Net-Energy-Metering-in-Hawaii.

Mostafa, M. (2014). *Feed-in Tariff Design for Wind and Solar: Cairo Case Study*. Workshop Program- Lifting Energy Subsidies: A Pathway to Renewables. Presented published by the Cairo Climate Talks. Retrieved from http://cairoclimatetalks.net/sites/default/files/Marwa%20Mostafa.pdf.

NCCETC (North Carolina Clean Energy Technology Center). (2018a). *Feed-in-Tariff*. Database of State Incentives for Renewables & Efficiency: North Carolina State University. Retrieved from http://programs.dsireusa.org/system/program/detail/5671.

NCCETC (North Carolina Clean Energy Technology Center). (2018b). *Distributed Generation Tariffs*. Database of State Incentives for Renewables & Efficiency. North Carolina State University. Retrieved from http://programs.dsireusa.org/system/program/detail/596.

NCCETC (North Carolina Clean Energy Technology Center). (2018c). *Solar and Wind Energy Credit (Personal)*. Database of State Incentives for Renewables & Efficiency. North Carolina State University. Retrieved from http://programs.dsireusa.org/system/program/detail/50.

Nolden, C. (2012). *Regulating Energy Security Through the Diffusion of Innovation – A Community Perspective*. Exeter, UK, College of Life and Environmental Sciences, University of Exeter. Retrieved from www.exeter.ac.uk/energysecurity/documents/ese_resources_May2012/Colin_Nolden.pdf.

NREL (National Renewable Energy Laboratory). (2018). *Celebrating 10 Years of Success Hawaii Clean Energy Initiative 2008–2018*. United States Department of Energy. Retrieved from www.nrel.gov/docs/fy18osti/70709.pdf.

NREL (National Renewable Energy Laboratory). (Undated). *Renewable Portfolio Standards*. United States Department of Energy. Retrieved from www.nrel.gov/state-local-tribal/basics-portfolio-standards.html.

Page, S. (2015, May 7). Hawaii Will Soon Get All of Its Electricity from Renewable Sources. *Think Progress*. Retrieved from http://thinkprogress.org/climate/2015/05/07/3656346/hawaiis-green-grid-plans/.

Paidipati, J. & Romano, A. (2017). *Net Energy Metering and Rate Changes – Implications for Distributed Generation*. Presentation published by the United States Department of Energy. Retrieved from https://betterbuildingssolutioncenter.energy.gov/sites/default/files/Net_Energy_Metering-Rate_Changes.pdf.

Peterson, H. (2017, September 20). Solar Payback. *Solar-Estimate*. Retrieved from www.solar-estimate.org/news/2017-09-20-solar-payback.

Pintz, W. S. & Morita, H. (2017). Hawaii Policy Background. In *Clean Energy from the Earth, Wind and Sun: Learning from Hawaii's Search for a Renewable Energy Strategy* (1st ed.) (pp. 15–34). Cham, Switzerland: Springer International Publishing. Retrieved from https://doi.org/10.1007/978-3-319-48677-2.

Piwko, R., Roose, L., Orwig, K., Matsuura, M., Corbus, D. & Schuerger, M. (2012). *Hawaii Solar Integration Study: Solar Modeling Developments and Study Results*. Paper presented at the 2nd Annual International Workshop on Integration of Solar Power into Power Systems Conference, Lisbon, Portugal. Retrieved from www.nrel.gov/docs/fy13osti/56311.pdf.

Poullikkas, A., Kourtis, G. & Hadjipaschaalis, I. (2013). A Review of Net Metering Mechanism for Electricity Renewable Energy Sources. *International Journal of Energy and the Environment*, 4(6): 975–1002. Retrieved from www.ijee.ieefoundation.org/vol.4/issue6/IJEE_06_v4n6.pdf.

Price, S., Horii, B., King, M., Benedictis, A., Kahn-Lang, J., Pickrell, K., Haley, B., Kadish, J., Jones, R., Sohnen, J. & Bowers, J. (2013). Introduction. In G. Petlin & K. Wu (Eds.), *California Net Energy Metering Ratepayer Impacts Evaluation* (pp. 13–23). San Francisco, CA, United States of America: Energy Division, California Public Utilities Commission and Energy and Environmental Economics Incorporated. Retrieved from www.cpuc.ca.gov/WorkArea/DownloadAsset.aspx?id=4292.

PUC (Public Utilities Commission). (2010). *Annual Report: Fiscal Year 2009–10*. Honolulu, State of Hawaii. Retrieved from https://puc.hawaii.gov/wp-content/uploads/2013/04/PUC-Annual-Report-Fiscal-Year-2009-10.pdf.

PUC (Public Utilities Commission). (2013). *Report to the 2014 Legislature on the Public Utilities Commission Review of Hawaii's Renewable Portfolio Standards*. Honolulu, State of Hawaii. Retrieved from http://puc.hawaii.gov/wp-content/uploads/2013/04/2013-PUC-RPS-Report_FINAL-w-Appnds.pdf.

PUC (Public Utilities Commission). (2015). *Annual Report: Fiscal Year 2014*. Honolulu, State of Hawaii. Retrieved from http://puc.hawaii.gov/wp-content/uploads/2013/04/PUC-FY-2014-Annual-Report.pdf.

PUC (Public Utilities Commission). (2018). *Report to the 2019 Legislature on Hawaii's Renewable Portfolio Standards. Issues Pursuant to Section 269–95(5). Hawaii Revised Statutes*. Honolulu, State of Hawaii. Retrieved from https://puc.hawaii.gov/wp-content/uploads/2018/12/RPS-2018-Legislative-Report_FINAL.pdf.

Pyrgou, A., Kylili, A. & Fokaides, P. A. (2016). The Future of the Feed-in Tariff (FiT) Scheme in Europe: The Case of Photovoltaics. *Energy Policy*, 95: 94–102. Retrieved from https://ezproxy-prd.bodleian.ox.ac.uk:4563/10.1016/j.enpol.2016.04.048.

R. P. Delio and Company. (2012). *Solar PV Policy in Hawaii: Lessons from the Mainland*. Honolulu, State of Hawaii. Retrieved from www.rpdelio.com/sites/default/files/documents/PV-Legislation-in-Hawaii-compared-to-Mainland_wp_091412%5B2%5D.pdf.

Ramírez, F. J., Honrubia-Escribano, A., Gómez-Lázaro, E. & Pham, D. T. (2017). Combining Feed-in tariffs and Net-metering Schemes to Balance Development in Adoption of Photovoltaic Energy: Comparative Economic Assessment and Policy Implications for European Countries. Energy Policy, 102: 440–452. Retrieved from https://doi.org/10.1016/j.enpol.2016.12.040.

READ (Research and Economic Analysis Division). (2014). *State of Hawaii Electricity Generation and Consumption in 2013 and Recent Trends.* Honolulu, State of Hawaii: Hawaii Economic Issues. Department of Business, Economic Development and Tourism. Retrieved from http://files.hawaii.gov/dbedt/economic/data_reports/reports-studies/ElectricityTrendsReport2014.pdf.

READ (Research and Economic Analysis Division). (2015). *Solar PV Installations in Honolulu – An Analysis Based on Building Permit Data.* Honolulu, State of Hawaii: Department of Business, Economic Development and Tourism. Retrieved from http://files.hawaii.gov/dbedt/economic/data_reports/briefs/Analysis_of_Solar_PV_Activities_in_Honolulu.pdf.

Renewables 100 Policy Institute. (2015). *100% Renewable Portfolio Standard for Hawaii – Talking Points.* Retrieved from www.go100percent.org/cms/fileadmin/media_upload/projects_files/usa/hawaii_pdf/100__RPS_Hawaii_Talking_Points.pdf.

RMI (Rocky Mountain Institute). (2008). *Policy Recommendations for Hawaii's Energy Future.* Boulder, CO. Retrieved from www.hawaiicleanenergyinitiative.org/storage/pdfs/Policy%20Recommendations%20for%20Hawaii%27s%20Energy%20Future.pdf.

Savenije, D. (2015, May 6). Hawaii Legislature Sets 100% Renewable Portfolio Standard by 2045. *Utility Dive.* Retrieved from www.utilitydive.com/news/hawaii-legislature-sets-100-renewable-portfolio-standard-by-2045/394804/.

Shupe, J. W. (1982). Energy Self-Sufficiency for Hawaii. *Science, 216*(4551): 1193–1199. Retrieved from https://doi.org/10.1126/science.216.4551.1193.

Sillaman, J. (2015, August 11). How Much Does Electricity Cost in Hawaii? *Hawaii Life.* Retrieved from www.hawaiilife.com/blog/how-much-does-electricity-cost/.

Sinha, A. & De, M. (2016). *Load Shifting Technique for Reduction of Peak Generation Capacity Requirement in Smart Grid.* Paper presented at the 2016 IEEE 1st International Conference on Power Electronics, Intelligent Control and Energy Systems, Delhi, India. Retrieved from https://doi.org/10.1109/ICPEICES.2016.7853528.

SOEST (School of Ocean and Earth Sciences and Technology). (2013). Hawai'i Natural Energy Institute. In *School of Ocean and Earth Sciences and Technology 2012 Self Study* (pp. 173–198). Honolulu, State of Hawaii: Hawaii Natural Energy Institute, School of Ocean and Earth Sciences and Technology, University of Hawaii. Retrieved from www.soest.hawaii.edu/soest_web/Documents/SOEST_2012_self-study_chapters/SOEST_2012_SS_HNEI.pdf.

State of Hawaii. (2019a). *About the Hawaii Clean Energy Initiative.* State of Hawaii: Hawaii Clean Energy Initiative. Retrieved from www.hawaiicleanenergyinitiative.org/about-the-hawaii-clean-energy-initiative/.

State of Hawaii. (2019b). *DCA Overview/Services.* State of Hawaii: Consumer Advocacy – Public Utilities (DCA), Department of Commerce and Consumer Affairs. Retrieved from http://cca.hawaii.gov/dca/about/.

Stockton, K. M. (2004). Utility-scale Wind on Islands: An Economic Feasibility Study of Ilio Point, Hawai'i. *Renewable Energy, 26*(6): 949–960. Retrieved from https://doi.org/10.1016/j.renene.2003.09.015.

Stoutenborough, J. W. & Beverlin, M. (2008). Encouraging Pollution-Free Energy: The Diffusion of State Net Metering Policies. *Social Science Quarterly, 89*(5): 1230–1251. Retrieved from https://doi.org/10.1111/j.1540-6237.2008.00571.x.

Watts, D., Valdés, M.F., Jara D. & Watson, A. (2015). Potential residential PV Development in Chile: The Effect of Net Metering and Net Billing Schemes for Grid-connected PV Systems. *Renewable and Sustainable Energy Reviews, 41*: 1037–1051. Retrieved from https://doi.org/10.1016/j.rser.2014.07.201.

WEC (World Energy Council). (2016). *World Energy Resources – Solar 2016*. (pp. 26–27). London. Retrieved from www.worldenergy.org/assets/images/imported/2016/10/World-Energy-Resources-Full-report-2016.10.03.pdf.

Wiser, R., Barbose, G., Bird, L., Bolinger, M., Churchill, S., Cory, K., Deyette, J., Fink, S., Holt, E. & Porter, K. (2008). *Renewable Portfolio Standards in the United States: A Status Report with Data Through 2007*. Berkeley, CA: Lawrence Berkeley National Laboratory. Retrieved from https://doi.org/10.2172/927151.

Wiser, R., Namovicz, C., Gielecki, M. & Smith, R. (2007). The Experience with Renewable Portfolio Standards in the United States. *The Electricity Journal, 20*(4): 8–20. Retrieved from https://doi.org/10.1016/j.tej.2007.03.009.

Wood, L. V. (2016, June 13). Why Net Energy Metering Results in a Subsidy: The Elephant in the Room. *Brookings*. Retrieved from www.brookings.edu/opinions/why-net-energy-metering-results-in-a-subsidy-the-elephant-in-the-room/.

Wüstenhagen, R., Wolsink, M. & Burer, M. J. (2007). Social Acceptance of Renewable Energy Innovation: An Introduction to the Concept. *Energy Policy, 35*(5): 2683–2691. Retrieved from https://doi.org/10.1016/j.enpol.2006.12.001.

Yailen, J., Dickerson, C. A., Takanish, W. & Cole, J. (2012). Saying Mahalo to Solar Savings: A Billing Analysis of Solar Water Heaters in Hawaii. In the *2012 ACEEE Summer Study on Energy Efficiency in Buildings* (pp. 341–348). Washington, DC: American Council for an Energy-Efficient Economy. Retrieved from www.aceee.org/files/proceedings/2012/start.htm.

Yalcintas, M. & Kaya, A. (2009). Conservation vs. Renewable Energy: Case Studies from Hawaii. *Energy Policy, 37*(8): 3268–3273. Retrieved from https://doi.org/10.1016/j.enpol.2009.04.029.

Yin, H. & Powers, N. (2010). Do States Renewable Portfolio Standards Encourage In-state Renewable Energy Generation? *Energy Policy, 38*(2): 1140–1149. Retrieved from https://doi.org/10.1016/j.enpol.2009.10.067.

Zhai, P. (2013). Analyzing Solar Energy Policies Using a Three-tier Model: A Case Study of Photovoltaics Adoption in Arizona, United States. *Renewable Energy, 57*: 317–322. Retrieved from https://doi.org/10.1016/j.renene.2013.01.058.

Part IV
Mainstreaming solar energy in small, tropical islands

8 Conclusions on mainstreaming solar energy

Energy transitions and mainstreaming

This chapter concludes the book's journey along an energy transitions narrative that began with the use of energy resources such as ethanol, charcoal, biomass and metabolic energy; followed by oil, natural gas and electricity; and then, solar energy. 'Culture' and 'policy' were the two elements of energy transitions of the MLP that were in the limelight throughout the book, and as the title suggests, mainstreaming has implications for both and vice versa.

The MLP illustrates the vertical structure of energy transitions. The socio-technical regime is the mainstream structure of the MLP, and mainstreaming refers to a niche-innovation evolving out of its niche(s) and becoming part of the regime by being more widely adopted. Considering the content of Chapter 4 through to 7, and the nature of mainstreaming in the MLP, there are several qualitative tenets that characterize mainstreaming based on the scope of the research which informed this book. These can be arranged as pre-requisites for, facilitators of, and pseudo-indicators of mainstreaming.

Pre-requisites for mainstreaming

Mainstreaming:

1) is linked to the adoption and use of technologies.
2) needs adopters to have a basic awareness of the technology being adopted, its service(s), and the resource(s) powering it.
3) requires reducing uncertainty about the technology being adopted and satisfying a need for more detailed pre-adoption information.

Facilitators of mainstreaming

During mainstreaming:

1) certain properties make a technology more easily adopted – cost-effectiveness is one of the most important.

2) a technology is more readily adopted when it satisfies the culturally valued motivators for adopting it.
3) a technology is more readily adopted when it fits non-disruptively into adopters' lifestyles.
4) Policies can improve the adoption rate of a technology.

Pseudo-indicators of mainstreaming

Once a technology has been mainstreamed:

1) there will be varying degrees of resource and technological substitutions.
2) there will be new practices and norms emerging in the mainstream energy culture because the technology becomes material culture.
3) there will be multiple adoption pathways established which means that the technology's original adoption niche is outgrown.
4) there will be new actor-institutions introduced into the regime's institutional multi-actor network accompanied by new inter-actor relationships as well as actor-agency and actor-power changes.
5) there can be industrial, policy, cultural, scientific, market and technological transitions.
6) there will be systems-level feedback linked to the widespread adoption of the technology.

Mainstreaming could be potentially quantified but quantifying energy transitions is hard so absolutely quantifying mainstreaming is also expected to be difficult. Several elements are also too fluid to be readily quantified and one is 'culture'. This is because quantification implies that there is a threshold and once it is crossed, an innovation can be branded as 'mainstream'.

Nevertheless, understanding the role(s) of 'culture' in energy transitions will help spotlight its role in mainstreaming and will subsequently help tease out its interactions with policies and the associated implications during the mainstreaming of solar energy.

'Culture', 'policy' and energy transitions

The book showed that 'culture' has proactive and reactive roles during energy transitions. Based on the study conducted using energy cultures to define 'culture', these roles are specifically expressed through niche-creation, technological selectivity, resistance to changes and lock-ins, reactions to changes and adaptive capacity – all of which transcend the three planes of the MLP.

Niche-creation

Based on the samples of the mainstream energy cultures in Trinidad and Barbados, the material culture, norms, practices and external influences are electric-oriented. In addition, there is seemingly some consensus surrounding perceived household peak energy demand occurring in the night due to the external influence of workplaces and schools on householders' presences in their homes, for instance.

This means that there are techno-cultural niches for electric-oriented innovations that will not hamper the continued use and perceived quality of electricity required by households during the incumbent perceived peak times. These niches are not necessarily specific to PV and are also very broad. For instance, at the systems-level, integrating PV can perpetuate and in some cases exacerbate the technical issues surrounding balancing peak and off-peak supply and demand – this shows why the influences of the other elements of the regime, that is, markets, science, technologies, industries and policies, are important in niche-creation.

Examples of other features that can create and shape niches based on the energy cultures looked at in the book are climate change and global warming, international oil markets, finite fossil fuel reserves and the infinite energy supply from the sun.

Technological selectivity

Niche-creation coupled with the inherent norms related to innovations that have yet to be adopted can favour one technology over another. For example, based on the sample of the mainstream energy culture in Trinidad, the norms affiliated with the understanding of solar energy largely refer to PV, for example solar panels, batteries and inverters, as well as its relative advantages over the incumbent electricity from the utility company. So, PV may be more easily assimilated than a SWH.

In the wider technological sense, if a series of innovations which perform similar functions are presented to adopters (e.g. different brands of solar panels), then the ones that are more aligned with the inherent norms related to the motivations for adopting it (e.g. cost-effectiveness and convenience) will be adopted first – which makes these the more competitive.

Resistance to changes and lock-ins

Adopting technological innovations involves becoming aware of the innovation and gaining knowledge about it. If a potential adopter's specific information needs are not met, one reason why the innovation may not have been adopted just yet is because there is no basis to act on any newfound awareness and information, that is, no motivation. This marks a form of passive resistance to change especially considering the communication channels that may be involved, that is, active versus passive information gathering.

In terms of lock-ins, the electric-oriented niche-creation and technological selectivity mentioned earlier will stand provided that electricity retains its technical dominance as the energy carrier. Its technical lock-in is reflected culturally and is further reinforced by other forces such as electric appliance manufacturing for instance.

Another example of a reinforcing influence is the temporal lock-in of external influences like workplaces and schools on energy users' presences in homes and their routines. Whilst this can help create broad niches as shown before, it can also support a lock-in especially when shifting peak energy demand and behaviour change are considered with the use of PV for example.

Reactions to changes

Energy cultures can respond to change-influences such that new norms, practices, material culture and external influences emerge after the adoption of an innovation – in which case energy cultures are likely to increase in complexity after an energy transition. Energy cultures can also react by forming subcultures based on clusters of defining norms, material culture and practices.

For example, the sample of the mainstream energy culture in Barbados showed that residents developed the practice of switching their SWHs' electrical backup switches on and off to supplement the primary solar energy used for heating the water in the system. This is driven by the external influence of the weather (namely the cloud cover), and associated norms are the belief that switching to the electrical backup increases the electricity bill and vice versa for instance.

Adaptive capacity

This is the ability to adapt to changing conditions in the system. An energy culture can adapt to an energy transition's changes by developing new norms, practices, material culture and external influences; resisting to and/or moderation of changes due to the lock-ins from routines and habits; and guiding changes by creating techno-cultural niches that favour certain technologies over others. This adaptive capacity therefore plays on and involves the previous four roles highlighted.

'Culture' and 'policy' as interacting elements

With an appreciation for the roles of 'culture' in energy transitions and mainstreaming, the book's research has found four ways in which energy cultures and residential solar energy policies can interact in small, tropical islands which have yet to mainstream solar energy: policy niche-creation, policy selection, policy goal-setting and policy designing.

Creating niches for policies

The results from Chapters 5 and 6 showed that the mainstream energy cultures are electricity-oriented. The norms, practices, material culture and external influences relate to the use and costs of using electricity from the utility companies. This suggests that there are very broad techno-cultural niches for technologies that generate electricity.

Distributed residential PV can fit into these niches (instead of solar water heating). This is supported by solar energy norms which largely relate to PV, and the fact that water heating is not as culturally salient as its technical consumption is. The cultural consensus that exists with respect to the norms surrounding the perception that residents use the most energy during the evening suggests that broad techno-cultural niches further exist for PV designs that do not hamper the continued use of electricity when it is needed most, for example FiT-supported PV.

It is worth stating that battery-integrated options can fit into these niches but are likely to enter the regime after grid-connected PV as shown in Barbados and Oʻahu. Key reasons are that though battery storage costs are falling, they are still relatively expensive, and battery-integrated systems are more complex innovations than grid-connected PV.

The techno-cultural niches described here are very broad and not necessarily solar specific. This is because 'culture' is not a standalone element in niche-creation – if it was then PV would have already been more widespread in Trinidad for example. This illustrates the interconnectedness of the regime's elements from Chapter 2 and why other elements such as policies will be instrumental in mainstreaming distributed residential solar energy. Nevertheless, the creation of these broad techno-cultural niches shows that the incumbent energy cultures looked at have long-term cultural structures built around electricity as the dominant energy carrier.

Broad renewable energy ambitions, for example RPSs and RETs can support such techno-cultural niches because they create the broad institutional and policy space for renewable resources which complements the cultural elements creating the niches in question. Such policies can capitalize on the sustainability-related motivators for adopting solar energy because such targets offer quantitative references which PV adopters can feel a sense of ownership towards. They also play on the norms surrounding solar energy's relative advantages related to the sun's longevity versus the limited fossil fuel reserves.

Selecting policies

Capital cost and performance-based incentives

As outlined earlier, renewable energy goals can support the broad niches for distributed residential solar energy. However, they do not directly address cost-effectiveness at the household level. This is important because people value 'costs' and are especially motivated by 'cost-effectiveness'. This will also be important in islands like Trinidad where the electricity is subsidized. These notions are related to the belief that using PV can be more economic than the electricity provided by the utility company on the island and so point towards a need for fiscal interventions. Such policies can be capital cost-oriented or performance-based.

Several of the residents interviewed in Barbados claimed tax credits on their SWHs which reduced the capital cost of the systems. But the policy was not an organic or salient part of their energy culture as shown by the discussion in Chapter 6. A more culturally impactful policy would arguably be a grid-connected programme like a FiT or NEM for example. Policies like these are examples of performance-based incentives.

These are likely to be more culturally impactful because they have longer lifetimes compared to one-off tax incentives. Policies are external influences on energy cultures and grid-connected incentives are likely to particularly affect the 'cost' aspects of residents' perceptions of household energy. This does not mean that addressing capital costs is unimportant, however.

The residents' norms and practices involved in adopting SWHs (see Chapter 6) in Barbados show that it is important to have policies which support flexible payment schemes in addition to Government-led policies, for example rent-to-own plans from retailers. These will be important for widening the adoption audience since flexible capital cost payment plans do not necessarily reduce the capital costs of solar systems, but they mitigate the impact of their upfront costs and make payments more manageable.

The residents interviewed in Trinidad value convenience and practicality amongst others as adoption motivators. So, fiscal policies should be designed around such motivators as well. For example, rent-to-own plans which mitigate the full impact of upfront capital costs will be more practical for many adopters in terms of making payments versus paying the full cost at one time. This also applies to how these payments are made where automated payments from a customer's bank account to the retailer through a mobile app may be more convenient than personally visiting the retailer or bank to make regular payments for instance. The administrative elements of accessing incentives like tax incentives, FiTs, NEM, flexible payments, and any others to be implemented, for example completing and submitting application forms, also need to be tailored to the culturally salient adoption motivators.

Technology and industry standards and certification and accreditation schemes

The salient information residents would want on household solar energy technologies and residents' adoption motivators can be used to design technology and industry standards. The norms that led to the number of adoption pathways observed for solar hot water in Barbados, for example non-manufacturer retail, door-to-door sales and direct manufacturer retail (see Chapter 6), show that there should be a multiplicity of adoption pathways. Having standards will ensure that the technology and retail service quality is maintained especially where non-manufacturer retailing is allowed (considering the risk of inferior systems and services entering the market).

Industry standards should be supported by certification and accreditation schemes. These schemes would ensure that retailers are skilled enough so that a minimum acceptable service standard guided by national industry and technology standards can be set. Other benefits that can come from certification and accreditation include publicity, branding and status for certified and accredited retailers which can be used for marketing their products. The certification and accreditation should further be supported by training and capacity building programmes to build a minimum and standardized skill-level in the industry.

Education and awareness-raising

In islands where there are no incentives available for household solar energy technologies, having education and awareness-raising initiatives will be particularly important because even in cases where there may be policies supporting solar energy, if adopters are unaware of them or the supported technologies then the policies' efficacy will be limited – as is the case in Trinidad.

Chapters 5 and 6 showed that the residents in Trinidad have a more generic understanding of solar water heating when compared to those in Barbados. Their understanding of it also includes references to PV and misconceptions about how both SWHs and PV work. This is not likely to be an isolated observation in islands so there is therefore a need to fill such knowledge gaps.

Key information that would help based on Trinidad's mainstream energy culture sample largely relate to solar energy technologies and the industry. For example, technological information will include brands and types of solar energy systems available as well as their lifetimes and reliability. Information about the industry particularly includes the installation information, maintenance regimen and standard warranties that should be offered for instance.

This awareness-raising also needs to consider the impact of policies because solar energy technologies are innovations, but their supporting policies will be innovations as well. For example, comparing standalone PV to PV under a grid-connected scheme are different innovations being presented to a potential adopter because the policy affects the technology's relative advantages and consequently the information presented the adopter. There will also likely be more uncertainties associated with policy-supported solar technologies because they are arguably more complex innovations.

Further, disseminating information should use the culturally salient information channels, for example the internet and one-on-one professional consultations; the internet is an example of a mass communication platform and the latter is an interpersonal channel. Mass communication tends to build awareness about innovations, but interpersonal channels are more effective in stimulating changes in attitudes and the decision-making associated with adopting innovations. Therefore, education and awareness-raising need multi-channel communication strategies to treat different types of adopters.

Nevertheless, education and awareness-raising policies will directly impact energy cultures by helping to culturally define the niches for solar energy described earlier on in the chapter and specifically so by influencing the norms about solar energy and the technologies.

Setting policy goals

As shown in Chapter 6, the development of new norms and practices, and acquisition of new material culture marks an energy cultures shift – and policies can directly and indirectly (through the supported technologies) catalyse such shifts.

The education and awareness-raising strategy outlined before is an example of a policy that will have a direct impact on an energy culture and tax credits are an example of a policy that will have an indirect impact because its assimilation in the energy culture will be strongly mediated by the PV or SWH systems they support.

A grid-connected policy like a FiT or NEM, however, is an example of a policy that will likely have both direct and indirect impacts on an energy culture. The anticipated direct and indirect impacts are based on comparing the cultural changes that would be likely to occur with the adoption of stand-alone PV versus grid-connected PV. This helps make the distinction between impacts due to adopting the technology versus adopting the policy and technology.

For example, in the case of a FiT, it will be an external influence on the energy culture. So as an external influence it would make solar panels new material culture (indirect impact); a new norm would be that adopters expect and believe that the amount of solar power generated affects their electricity bill (indirect impact); and a new practice would be collecting cheques/payments for generating and/or exporting solar power to the grid (direct impact) (see Table 8.1). These anticipated changes are part of the energy cultures shift under FiT-supported PV and can be framed as goals that policymakers can design the FiT to achieve.

Designing policies

A FiT is a good example to look at to illustrate how energy cultures and policies can influence each other. Based on the research conducted which informed the book, the considerations for the metering and use of a generation and/or export tariff under a FiT are worth highlighting as specific examples.

Solar power exported to the grid can be either estimated or metered. Chapter 5 touched on the perceived electricity bills of the residents interviewed in Trinidad as well as provided insight into their perceptions behind these bills; these perceptions determine whether householders believe their electricity bills are either too high or too low. Whilst there were no overwhelming consensuses driving the rationales, four stood out because of two

Conclusions on mainstreaming solar energy 195

Table 8.1 Showing key energy cultures policy goals that can be associated with a FiT

Policy goal	Goal type	Impact
To integrate the PV system (primarily solar panels) as material culture	Material Culture	Indirect
To integrate an added electricity generation meter as material culture	Material Culture	Direct
To promote the application for planning permits	Practices	Indirect
To promote the application for grid interconnection permits	Practices	Direct
To promote the adoption of PV systems	Practices	Indirect
To promote the adoption of a generation meter	Practices	Direct
To continue paying a household electricity bill	Practices	Indirect
To introduce collection of cheques/payments for the solar power exported and sold	Practices	Direct
To normalize the belief, aspiration and expectation that solar energy is being used in the home	Norms	Indirect
To normalize the belief, aspiration and expectation that using solar energy reduces the costs of electricity	Norms	Indirect
To normalize the belief, aspiration and expectation that using solar energy reduces the consumption of electricity	Norms	Indirect
To acknowledge that the weather impacts the household's solar energy generation as an external influence	External Influence	Indirect
To acknowledge that the contracted residential solar tariff rate determines the household's costs of electricity as an external influence	External Influence	Direct

Source: The author.

Note
The direct and indirect impacts are based on comparing the changes that would be likely to occur with stand-alone PV versus distributed PV.

underlying beliefs: the electric utility company determines the costs on bills by zoning households based on an urban-rural divide, and the electricity bill is calculated remotely versus reading the household's actual electricity meter.

Whilst these are residents' perceptions and not statements of fact, they still point towards there being a belief that the utility's metering is flawed and there is mistrust in the billing. Therefore, this may affect the understanding of the metering and crediting/debiting for FiT-supported PV. So, having a separate generation and consumption meter (versus a bi-directional meter as with NEM for instance) may be a better fit here for example. Further, Chapter 7 showed that if solar power is metered under a FiT, it would use separate generation and consumption meters because the imported and exported electricity is priced differently.

Chapter 6 showed that consuming solar energy in the home is a norm for the residents interviewed in Barbados, that is, solar energy self-consumption. This is important because it diversifies the household energy mix (technically and culturally) and gives 'the household' a greater sense of ownership over their energy (re)sources. Encouraging solar self-consumption can be a cultural policy goal (see Table 8.1). This goal helps with the considerations around a FiT's generation/export tariff design since different designs encourage self-consumption.

As Chapter 7 outlined, in the most polarized of form, homes can be paid either for generating their solar power and/or exporting it – and if the export tariff rate is higher than the utility's retail rate then exporting power is more profitable, but when the export tariff is lower it encourages self-consumption. This suggests that a solar power generation tariff, or export tariff that is lower than the utility's retail rates, would encourage solar power self-consumption.

Using a generation tariff means that any solar power exported to the grid is done freely and the utility would benefit from essentially 'free' power. Therefore, solar power self-consumption is in the household's best interest. But whilst a generation tariff encourages self-consumption, there is the risk of adopters increasing their energy consumption to avoid exporting it freely if there is a surplus, for example leaving the air conditioning on for longer periods of time.

Householders may also not care whether they export power freely to the grid since they are being paid for their solar power no matter where it is consumed. This is because the norms which may develop around the FiT may include a degree of insensitivity to differences between generation and export tariffs – the fact that adopters are simply offered generic payments for their solar power may be the way in which the policy's 'incentivization' is culturally assimilated.

In order to reconcile solar power self-consumption with the incentive to adopt PV under an exportation scheme, the export tariff should be geared towards surplus solar power generation and remunerated at a rate lower than the retail electricity tariff. So, the fact that adopters are paid for surplus export means that they need to consume their solar power and there is an incentive to reduce consumption so that more power could be exported. This allows the home to consume their solar power, reduce energy consumption and still benefit from the solar power payments.

These considerations do not exclusively point towards choosing between a generation and export-oriented FiT for surplus power or even combinations of the two. But they do however put forward considerations that would be important for planning a policy like a FiT from a more human angle which is rooted in energy cultures.

An indirect aspect of the FiT design would be advisories to residents on how to get the most out of their PV system under a FiT. This specifically refers to the disparity between peak solar power generation and peak energy demand raised in Chapter 7. This warrants thinking about demand-side management interventions aimed at load shifting to encourage solar power self-consumption during peak solar power generation times.

The book's cultural chapters provided detail on the perceived peak energy demand timings and their drivers, but a significant aspect is the fact that

Conclusions on mainstreaming solar energy 197

householders are simply not home during the day which is when solar power generation is at its peak. This means that there will be more solar power exported then since it is not being consumed and generation is at its highest.

Practices such as cooking and showering, and material culture like televisions and air conditioners are more specific drivers of the perceived evening peak demand. Specific and flexible features of energy cultures can be identified and used as targets for behavioural change through load shifting. For example, doing the laundry is a salient driver of the perceived peak demand in Trinidad and Barbados, and weekends are also part of the residents' cultural framing of a 'typical' day. So, households can be encouraged to do their laundry during peak solar energy generation hours on weekends if the laundry is being done manually. Alternatively, if laundry machines are 'smart', then they can be automated to wash clothes during peak solar power generation times.

Mainstreaming solar energy: the cultural and policy implications

So, having characterized mainstreaming, presented the roles of 'culture' in energy transitions and most recently illustrated how policies and energy cultures can interact during mainstreaming, the last aspects that still need to be highlighted are, as the book's title suggests, the cultural and policy implications linked to mainstreaming solar energy.

For small, tropical islands considering mainstreaming solar energy technologies like SWHs and PV into their residential electricity regimes, it should be understood that:

1) mainstreaming the technologies will make the regime's multi-actor network and the mainstream energy culture more complex.
2) policies are not only economic or political instruments; policies are external influences on energy cultures because they are outside of the direct influence of the energy cultures' members.
3) the nature of the cultural information which aggregates around households' energy costs, that is, energy costs as a cultural domain, suggests that it is most sensitive to external influences when compared to that of energy supply or even energy usage.
4) arguably the most impactful policies would affect households' energy usage and its costs – and this would be reflected in the nature of the cultural information associated with these respective cultural domains which are attributable to the policy's implementation.
5) policies can directly and indirectly (through their supported technologies) catalyse cultural transitions, and these shifts will be dependent on the policy's design.
6) policies that affect the timing of when householders' practices are performed (particularly routinized practices) will be disruptive because it

will be trying to break a lock-in that is not completely dependent on in-house activities since there are external influences which will affect the timing of these practices.

7) modern mainstream energy cultures are predominantly electricity-oriented and have norms and practices linked to the perceived peak consumption being in the evening/night. This suggests that there is a broad niche for solar energy policies that support the continued use of electricity at this time of day without disruption to a given household's electricity supply and the quality of services it affords during these times.

8) technological transitions unfold in energy cultures. This suggests that policies should capitalize on their associated technologies' relative advantages since the incumbent technologies will be put up against the incoming solar technologies.

9) any adjustments made to a utility company's cost recovery such that the residential retail tariff is affected (and particularly in cases where subsidy removals are involved) are an external influence which consequently affects the norms and practices surrounding energy usage and its costs.

10) policies such as RPSs or RETs encourage a stipulated renewable energy contribution in any given energy mix. Such policies play on the motivation for adopting solar energy because it is a sustainable energy resource as well as existing norms surrounding the relative advantages of solar energy.

11) policies such as NEM and FiTs allow PV owners to generate, consume and export and sell their solar power to the utility company. In the case of the latter, making use of generation and/or export tariff can support solar power self-consumption by households to varying degrees. NEM and a FiT design that encourages customers to use their solar electricity in the home, that is, solar power self-consumption makes solar energy a household resource and a norm to use it.

    i   adopting household PV under NEM or a FiT will develop new norms, practices and material culture.
    ii   the rates provided in NEM or a FiT's long-term contracts will affect the perceptions of household energy costs because it is an external influence on energy usage and its costs.

12) there are misconceptions and knowledge gaps existing about solar energy and solar energy technologies. This calls for policy solutions that include education and awareness-raising campaigns.

13) the potential for saving money is the most salient relative advantage of solar energy over the incumbent electricity provided by utility companies. 'Cost' is the most salient pre-adoption information desired and 'cost-effectiveness' is the most important adoption motivator. Therefore, fiscal policies that incentivize solar energy can be culturally effective.

14) adopters have several motivations for adopting a technology. These should be important considerations for policies to set industry and technology standards.
15) policies should encourage a multiplicity of adoption pathways that are guided by best-practice and lessons-learnt guides, industry and technology standards and certification schemes.
16) policies should encourage flexible payment plans for capital cost payments.

Critiquing the research behind the book

Critiquing the value of studying 'culture' in energy transitions

Having come to the end of this book, you should now have a good grasp of what mainstreaming means beyond the conventional sense and how 'culture' evolves in energy transitions, as well as be able to see policies as more than just socio-economic instruments, but also facilitators of cultural change. But the research and narrative are not without their limitations.

Theory

'Behaviour' is a fundamental element of the 'socio' in socio-technical studies and 'culture' is one way to approach behaviours. But 'culture' is so complex that it transcends the planes of the MLP, so it is vertically structured relative to the MLP. As a result, this is a particularly useful way to approach 'behaviour' since it means that its roles in socio-technical energy transitions can be studied holistically.

More critically, in terms of the explicit inclusion of 'culture' in the MLP, it seems to be a term used to represent the social and more human elements of the structure. Therefore since 'culture' is but one expression of this, it is hard to grasp conceptually, and as 'behaviour' is at its root, 'culture' could be potentially replaced with 'behaviour' in terms of terminology.

However, when considering the element of time and the systems-level context of the regime, 'culture' is a better concept because it acknowledges a collective system of evolutionary and integrated behaviours. The ECF was consequently useful for illustrating this through the interactions of norms, practices and material culture. Its consideration for an external context is also ideal because it essentially sets the energy culture and its behaviours in the wider context of the MLP.

Policy practice

Of the elements in the regime (see Chapter 2), the book focused on 'culture' and 'policy'. Despite the multiplicity of factors that the book shows can affect policymaking and implementation, there seems to be a common policy framework applied in the three case studies to support the mainstreaming of household solar energy technologies like SWHs and PV, that is:

- forms of awareness-raising, e.g. educational campaigns and retailers' marketing
- technology and retail certifications/accreditations/standards, e.g. wiring codes
- implementing capital cost reduction incentives, e.g. tax credits
- implementing a grid-connected scheme for distributed PV, e.g. a FiT, NEM or a contextual variant of performance-based incentivization
- adopting overall clean/renewable energy (including energy efficiency) ambits, e.g. RETs or RPSs.

A seemingly conventional policy outlook would brand the first of the above list as more cultural than the others – with the others being largely economic, legislative and market-oriented. In the case of the incentives, such policies are often framed as economic agents. It is because of this outlook that 'culture' has been undervalued in traditional policymaking (and to some extent disregarded). So, one can question whether 'culture' will be that influential especially if it supports a fiscal policy that could have been implemented without considering it. However, this assumes that there is a capacity for seamless policymaking being applied to rational economic actors.

Such perspectives also make it seem like narratives such as the one presented in the book boil down to 'policy' versus 'culture'. It is worth more to approach it as what value does studying 'culture' add to policymaking and implementation. Nevertheless, the book showed that the regime's elements transition at different rates over varying timescales. This implies that a regime's elements do not have equal importance or influence on an energy transition. 'Culture' may play an influential role in some geographies but less so in others. But this depends on how 'culture' and 'the regime' are defined relative to geography.

Additionally, using the selected case studies meant that the available suite of policies which could be analysed would be those implemented in the islands. So even though the various policy options looked at were used to mainstream solar energy, it does not mean that they are the most effective policies to be considered in other small, tropical islands. This also meant that the interactions between 'policy' and 'culture' were limited to the policies used in the study. So, there may be other policies that have more visible or influential relationships with 'culture' but may have been omitted because they were not implemented in Trinidad, Barbados or O'ahu.

Contributions of the book and opportunities

Considering the number of small, tropical islands in the world, the book's takeaways are highly applicable. More specifically, the book betters the understanding of the residential energy transitions occurring in Trinidad, Barbados and Oʻahu. It offers readers from other islands a chance to see where theirs would potentially fit in the narrative and what lessons are transferable.

The book improves the spatial understanding of the MLP by applying it to not one but three case studies and by being applied to small, tropical islands represents a highly specific geographic application. Small, tropical islands are also amongst the most vulnerable geographies relative to the repercussions of energy insecurity and climate change. So, this research on residential solar energy is a valuable contributor to understanding the mainstreaming of alternative energies which can mitigate the impacts of both in these marginal geographies.

Considering the book's contributions, there are several opportunities for further research on small-island energy cultures; below are two of the more interesting examples.

There is an opportunity to research both grid-connected and off-grid residential PV systems and their impact on energy cultures. In terms of grid-connected PV, given that SWHs were looked at in the book's cultural work and were also used to make policy-based deductions for PV, it is only logical that an explicit study be done to investigate the impact of supporting policies like FiTs, NEM or other location-specific examples such as CGS.

In the case of off-grid PV, this can look at battery-integrated designs or even standalone PV that directly power several household appliances. This would be an important project given that batteries are becoming prominent global socio-technical elements. Such research may even extend to the adoption of electric vehicles given that this is another globally emergent sociotechnical development and using solar power to charge the vehicles is a useful application to investigate as it pertains to load shifting.

Further, there is scope to do long-term studies on the cultural drift of the features that appear within a given energy culture. Such a study will provide insight into if there are feature-migrations due to changes in perceived salience; their entry and exit patterns; lag times in response to stimuli such as policies or technology adoption; and their rates of migration based on the saliences of the cultural features being observed over time.

Small, tropical islands provide well-defined geographic lenses for studying energy transitions and are one of the best contextual settings to illustrate how space, place and geography affect energy transitions and vice versa. Their land-sea borders are strong physical and conceptual limiters for their electricity systems – and each island has its own character (as shown in the book). But their smallness, remoteness and oceanic natures, that is, their insularity, make island energy research highly replicable and scalable.

You've come to the end a historical journey involving cultures, technologies and policies. But, do you remember when you first sat down on that white, warm beach with only the 'sun, sea and sand' for company in the first chapter? No? Then take a look at Figure 8.1 ...

Figure 8.1 Showing seagulls gliding in the sky along a tropical beach.

I'm sure the tropical breeze that's once again blowing and causing the coconut trees behind you to whisper to each other can remind you – so too that faint, yet oh so recognizable series of squawks from the seagulls casually gliding by overhead.

You've been here so long that the tide's already in, and you're now sitting in the cool, sandy waters of the waves gently going in and out behind you. Where'd the time go? Even the sun's no longer where it was in the sky when you first sat down – it's closer to meeting the horizon and yet to you ... yet to you it's light is still so warm ... still so radiant ... still so, quite frankly, 'tropical' ...

Looking back at what the book has taken you through, it's certainly a profound thing to think that all of it is linked to as simple a thing as the rays of sunlight still striking your face; as simple a moment as just sitting there feeling the sun's warmth is packed with potential for providing energy that can change and improve the lives of those who call small, tropical islands home.

Reference

pasja1000. (2018) [*Untitled illustration of seagulls gliding in the sky along a tropical beach*]. Pixabay. Retrieved from https://pixabay.com/photos/palm-trees-exotic-blue-sky-palm-3540107/.

Index

Page numbers in **bold** denote tables.

adaptation 5
adaptive capacity 32, 52, 188, 190
adoption 57, 108, 127, 128, 156, 192, 198–199; of electric vehicles 201; of innovations 7, 38, 58–60, 190; of photovoltaics and/or solar water heaters 9, 12–13, 36, 39, 49, 55, 83, 87, 128, 131–134, 141, 143, 145–146, 153–154, 157, 167, 194–**195**; of renewable energy 10, 167; of technologies 53, 57, 187–188, 201
agency 169, 173; in energy cultures 54; as an institution 78–79, 81, 86; in the regime 32, 188
alternative current (AC) 110
appliance (s) 97, 115, 201; damage 111; usage 99–107, 110, 113, 134, 137–138, 140, 190; *see also* material culture
assumption (s) 33, 39, 49, 55
awareness 95, 187, 189, 193–194, 198, 200; in and for the public 81–**82**, **85**, 87; of solar water heating technologies 114, 130, 131; *see also* knowledge

bagasse 69, 86, 125
Barbados 114–115, 191–192, 200–201; case study background 8–9, 11–13, 15, 17, 35–36, 39, 117, 125–127, 153, 155, 157, 173, 188; energy cultures and solar energy in 9, 127, 131–132, 134–135, 137–139, 141–147, 171, 193, 195, 197
Barbados Light & Power Company (BL&P) 125–126, 143–145, 147
Barbados National Oil Company (BNOC) 125
Barbados Renewable Energy Association (BREA) 126

barrels of oil 5, 11, 72–74, 125
barrels of oil equivalent (BOE) 10–11, 74, 125
barrier (s) 7, 17, 81–83, 87, 159
baseload 99–100, 105, 115
bi-directional meter 164, 172, 195
biomass 17, 28, 69, 94, 155, 187

capital costs 82, **127**, 129, 169; are high for PV 83, 166, 172; policies 169, 191–192, 199
carbon dioxide (CO_2) 10
Caroni 1975 Limited 70
certification 159, 192–193, 199
charcoal 70, 86, 187
climate change 111, 169; as an external influence 137, 147; and islands 5, 10, 201; and niche-creation 189; and policy 80
cloud cover 106, 138–139, 143, 147, 190
coal 28, 71, 155–**156**
colonization 4, 12, 69–70
complexity 171; and culture 49, 58, 116, 190; of the energy system 98; of energy transition 6, 40, 157
consumption meter 172, 195
cost recovery 100, 172, 198; and feed-in tariffs 168; and fuel costs 126, 138, 157
cultural change 7, 199; mainstream cultural change 125, 134; *see also* energy culture (s) shift (s)
cultural transition 142, 146; *see also* cultural change; energy culture (s) shift (s)
culture (s) 4, 6–7, 12–13, 15, 17–18, 53, 56, **82**, 87, 94–95, 99, 116, 141, 187–188, 190–191, 201; definition 49–51; in and geography 51–52, 56;

204 *Index*

culture (s) *continued*
  energy transitions 31, 37, 40, 52, 58–61, 197; theory 53, 128, 199–200; *see also* mainstream culture (s)
Customer Grid-Supply (CGS) 170, 173, 201
Customer Grid-Supply (CGS) Plus 170, 173
Customer Self-Supply (CSS) 165, 170–171, 173

degrees celsius (°C) 10–11, 13
Department of Planning and Permitting (DPP) 159
diffusion of innovations 30, 37
direct current (DC) 110; *see also* alternating current (AC)
distributed photovoltaics (PV) systems: in Barbados 17, 126, 143–145, 153; in Oʻahu 17, 155–157, 159, 163, 171; in Trinidad; 79, **84**
distribution network 79; and government 81; and household photovoltaics 145, 156; and transmission network **97**, 110, 126; and utility companies 75–76, 87, 98, 158
Division of Consumer Affairs (DCA) 159
draft animals 70, 86
dry season 106

economic transitions 75
economy 4, 28, 34–35, 37; in Barbados 12; Oʻahu 13; in Trinidad 10, 69–72, 74–75, 86
education 6, 81–**82**, **85**, 114, 193–194, 198; *see also* knowledge
electric water 103–105, 113, 116, **127**, 129–131, 143
Electrical Inspectorate Division (EID) 78–79
energy bills 111, 141
energy consumption 6, 14, 29, 34, 99, 105; by appliances 99–100, 116, 147; in households 103, 107, 139, 141, 159, 196; by society 165; *see also* energy demand (s); energy usage
Energy Cost Adjustment Clause (ECAC) 157
energy costs 5–6, 105–108, 111, 115, 141, 197–198
energy culture (s) 17, 49, 188–199, 201; in Barbados 125, 132, 134–135, 137, 140–143, 146; theory 53, 55–61; in Trinidad 94–95, 98, 100, 103, 108, 115–116; *see also* mainstream energy culture (s)

energy culture (s) shift (s) 57–58, 132, 194
Energy Cultures Framework (ECF) 17, 53–55, 57–61, 102, 199
energy demand (s) 103; and household perceptions of peak usage 100–103, 105, 115, 137, 146, 165, 189, 197; in Oʻahu 14; peak 99, 102, 110, 115, 137, 162, 165, 173, 196; shifting peak energy demand 165, 190; in Trinidad 11
energy efficiency **85**, 87, 154, 159, 162, 172
energy efficiency portfolio standards (EEPSs) 162–163, 172
energy infrastructure: centralized 6, 8, 31, 36, 37, 87, 94, 126–127, 143, 145, 155, 171–172; decentralized 6, 8, 36, **136**, 156, 172
energy input 95, **97–98**, 136, 142
energy insecurity 5, 201
energy storage: batteries 9, 39, 109, 134, 165, 170–171, 173, 189, 201; battery 15, 165, 171, 173, 191, 201
energy subculture (s) 56–59, 61; *see also* niche-level culture (s)
energy supply 17, 29, 35, 74, 86, 160, 189, 197; as a cultural domain in Barbados 134, 136, 142; in cultural domain in Trinidad 94–95, 98
energy transition (s) 2, 6–8, 13, 15–18, 28–30, 49, 127, 187–188, 197, 201; in Barbados 125, 132, 143; and culture 51–52, 58, 60–61, 188, 190, 199; in Oʻahu 158, 170, 172–173; in theory 30–35, 39–40, 95, 157; in Trinidad 67, 69, 75–75, 86–87, 103, 108
energy usage 98, 197–198; as a cultural domain in Barbados 136–140, 142; as a cultural domain in Trinidad 100, 103, 106, 115–116; *see also* energy demand (s)
enslavement: slave 70; slavery 12
Environmental Policy and Planning Division (EPPD) 78, 80
export arrangements: and the economy 5, 10; of hydrocarbons 70, 71–72, 74–75, 78, 82; of solar power 36, 144, 157, 164–165, 168, 170–172, 194–198
external influence (s) 94, 188–190, 192, 194–198; in energy cultures framework 53, 55–56, 58, 60–61; Trinidad examples 99–100, 102, 105–106, 115–116; Barbados examples 134, 136–139, 143, 146–147

Index 205

Fair Trading Commission (FTC) 126, 144–145
feed-in tariff (FiT) **84**, 87, 191–192, 194–196, 198, 200–201; in Oʻahu 167–173; and other similar policies 154, 161, 192
fossil fuel energy resources: in Barbados 125–127, 142–145; and energy transitions 6, 28–29, 34, 40; and islands 4–5, 17; in Oʻahu 154–155, 159–161; reserves 189, 91; in Trinidad 10, 69, 73, 80, 86–87, 110, 115
Fuel Cost Adjustment (FCA) 126, 138–139, 144–146, 157

generation meter 195
geography 4, 37, 51, 200–201
gigawatt-hours (GWh) 156, 162
gigawatts (GW) 8, 29
global warming 80, 111, 189
governance 94; and capitals 10, 13; and electricity 75, 126; and institutions 77–78, 81, 94, 159–160
government (s) 6, 10, 73, 147, 192; of Barbados 131, 143; governmental 4, **82**, 83, 85, 86, 126, 131; of Hawaiʻi 160, 166, 169, 172–173; of Trinidad 74–75, 77, 79–83, 85
greenhouse gas emissions 10, 110, 160
grid (s) infrastructure: network 4, 14, 31, 36, 77, 96, 102, 137, 144–145, 155–157, 160, 163; grid-connected 158, 162, 191, 201; grid-connected policies 36, 84, 164–166, 168, 170–173, 192–196, 200; plumbing 113
gross domestic product (GDP) 10, 12–13

Hawaiʻi 13–14, 36; energy resources in 153–156; energy usage and costs in 157–159; energy institutions in 160; policies in 161–165, 167–173; *see also* Oʻahu
Hawaii Clean Energy Initiative (HCEI) 161–162
Hawaii Electric Light Company (HELCO) 155, 168
Hawaii Natural Energy Institute (HNEI) 159–160, 162
Hawaii Solar Energy Association (HSEA) 159, 161
Hawaii State Energy Office (HSEO) 159–160, 173
Hawaii State Legislature (HSL) 159, 160, 173

Hawaiian Electric Company (HECO) 14, 155–**156**, 159–160, 163, 168, 170–171
Hawaiian Electric Industries (HEI) 14, 155, 163, 168, 172
hurricane (s) 5–6, 9, 11

import arrangements: in islands 4–6; in Barbados 11, 14, 125–126, 145, 147; in Hawaiʻi 154–155, 157–158, 160, 163, 171; solar power 157, 164–165, 195; in Trinidad 72, 75, **84**
incentive (s) 78, 86, 110, 170–171, 191–193, 196, 200; *see also* tax incentive (s)/credit (s) schemes
incoming solar energy: incident solar energy 138; incident solar radiation 8
indentureship 70
independent power producer (s) (IPP (s)) 36; in Barbados 126, 144; in Oʻahu 155–**156**, 159; in Trinidad 76–77, 87, 94
industrial transition (s) 86, 172
insularity 4, 35, 51–52, 201
internal rate of return 167
international oil price (s): 107, 126, 146, 157; and fluctuations 75, 107, 115, 131, 139, 144, 155, 172; and the landscape 73
inter-policy transitions 173
intra-policy transitions 172

kilometres (km) 9, 11, 13
kilowatt-hour per square metre (kWh/m^2) 8–9
kilowatt-hours (kWh) 126, 144, 162–163; household rates and consumption in Barbados 126; household rates and consumption in Oʻahu 157–159; household rates and consumption in Trinidad 77, 82, 107; and solar energy policies 164, 166, 168, 172
kilowatt-peak (kWp) 143–144
knowledge 103, 146, 189; as a part of culture 49; in socio-technical systems 29; and solar water heater adoption 108–109, 114, 127, 132, 193, 198

landscape 30–31, 33–35, 39–40; forces and culture 52, 55–56, 58, 60–61; examples 102, 115, 154
legislation 76, 153, 160, 163
lifestyle (s) 6, 128, 146, 188; in households 101–102, 105, 115
light-emitting diode (LED) 103, 116
lignite 71, 86

limitation (s) 33–34, 55, 199
liquefied natural gas (LNG) 74
liquefied petroleum gas (LPG) 35; in Barbados 125, 129, 134, 140–142; in Trinidad 70, 86, 95, 98, 104–105, 110, 115–116
lock-in 34, 55, 115, 189–190, 198

mainstream culture (s) 1, 51–53, 55, 58, 60–61, 125; *see also* mainstream energy culture
mainstream energy culture (s) 49, 87, 188–190, 197; in energy cultures framework 58–61; in Trinidad 94–95, 100, 103, 108, 115–117, 193; in Barbados 125, 132, 134–135, 137, 140–143, 146
mainstreaming 7–8, 18, 57, 73, 81, 83, 174, 185, 190, 201; and policies 83, 154, 172–173, 191; of solar energy technologies 15, 17, 39, 79, 87, 128, 134, 147, 157, 161, 200; theoretical definition 37–38, 40, 58, 61, 108, 132, 156, 187–188, 197, 199; *see also* the mainstream
market (s) 31, 38, 84, 153–154, 160, 167, 172–173, 200; hydrocarbon 5–6, 73, 75, 78, 82, 138–139, 144, 157, 169, 189; technology 7, 12–13, 33–35, 37–38, 83, 110, **127–128**, 131–134, 145–147, 156, 192; and energy transitions 34, 39–40, 61, 188–189
marketing 56, 131, 133–134, 193, 200
material culture 188, 190, 194–**195**, 197–198; in Barbados 134, 136–141, 146; and the theory 53, 55–61, 199; in Trinidad 94, 99–106, 115–116
Maui Electric Company (MECO) 14, 155, 168
megawatts (MW) 36, 162; generation capacity in Barbados 126; generation capacity in Oʻahu 155–**156**; generation capacity in Trinidad 11, 76; solar capacity in Barbados 144–145; solar capacity in Oʻahu 14, 168–169
metabolic energy 17, 28, 70, 86, 187
metering 108, 164, 194–195
millimetre (mm) 10–11, 13
Ministry of Energy and Energy Industries (MEEI) 78, 80
Ministry of Planning and Development (MPD) 78–79
Ministry of Public Utilities (MPU) 78–79
misconceptions 113, 114, 193, 198
mitigation 5, 111

motivations 36, 128, 131, 189, 199
Multi-lateral Environmental Agreements Unit (MEAU) 78, 80
Multi-level Perspective of Socio-technical Energy Transitions (MLP) 15, 17, 40, 102, 115, 201; and culture 52, 54–56, 58–61, 116, 188, 199; and mainstreaming 37–39; theoretical description 30–35

National Gas Company of Trinidad and Tobago (NGC) 74–76, 80, 87
National Petroleum Corporation (NPC) 125
natural gas 28, 35, 98, 187; in Barbados 11, 125–126, 145; in Trinidad 10, 17, 72, 74–75, 77, 80, 86–87, 94–95, **97**, 108, 111
net energy metering (NEM) 14, 192, 194–195, 198, 200; in Oʻahu 154, 157, 161, 164–167, 172–173; and other similar policies 169–171, 173
net energy metering (NEM) Plus 171, 173
network of actors: actor-network 30; actors 31, 54, 55, 60, 78, 160, 168, 171–173, 200; multi-actor network 32, 188, 197; *see also* agency
niche-creation 188–191
niche-innovations 30, 32–33, 39–40, 52, 58–59
niche-level culture (s) 52, 56, 58
norms 7, 188–192, 194–196, 198–199; and the mainstream 37–38; and culture 49, 52–53, 55–61; Trinidad examples 94–95, 99–100, 102, 105–106, 109, 111, 113, 115–116; Barbados examples 134, 136–43, 146–147

Oʻahu 35, 39, 115, 145, 147, 191, 200–201; case study background 8–9, 13–15, 17, 153–159; and solar energy policies 163, 165–168, 170–173
oil and gas 10–11, 13, 74–75, 77, 125
oil boom 72
oil well 71
Organization of the Petroleum Exporting Countries (OPEC) 72–73

paraffin 71, 86
path dependency 29–30
payback 166, 167
penetration 7, 132, 156
perception (s) 51, 53, 56; of household energy 101, 103, 107–108, 115–116, 134, 143, 191–192, 194–195, 198

petrochemical 72, 86
petroleum hydrocarbons: imports and prices 5, 14, 154–155, 157, 171; industry 70–72; legislation 73, 80
Petroleum Production Levy and Subsidy (PPLS) Act 73; *see also* subsidy
photovoltaics (PV) **136**, 143, 170
pitch 70–71, 86
policy design 190
policy oscillation 172
policy selection 190
policy transitions 171
policymaking 7, 10, 13, 160, 173, 199–200
population 4, 11, 13–14, 51–52, 56
power generation 31, 36, 40; in Barbados 126–127, 143–145; in Oʻahu 14, 155–156, 171–172; solar 165, 196–197; in Trinidad 11, 72, 75–77, 79, 81, **84**, 86–87, 95, **97–98**, 107, 110; *see also* peak solar power generation
Power Generation Company of Trinidad and Tobago (PowerGen) 76, 106
power purchase agreement (PPA) 76–77, 87
practice (s) 29, 94, 188, 190, 192, 194–**195**, 197–199; Barbados examples 134, 136–143, 146–147; in energy cultures framework 53–61; policy and codes of 79, 159, 199; Trinidad examples 99–106, 115–116
Public Utilities Commission (PUC) 159–160, 162–163, 167–168, 173

quantitative modifications 172

rain 10–11, 13, 106, 112, 141, 143
reacting to change 58, 188, 190
recession 154
regime (s) 189, 191; in theory 30–37, 39–40, 58–61, 187–188, 197, 199–200; in Barbados 125–127, 132, 143, 145–146; and mainstream culture 52, 54–56, 58; in Oʻahu 153–154, 157, 161, 170–173; in Trinidad 75, 80–82, 86–87, 94–95, 115–116
Regulated Industries Commission (RIC) 78–79, **84**
regulation 75, 77–78, 126, 159
renewable energy 7–8, 29, 142, 191; in islands 5–6, 81; in Barbados 145; in Oʻahu 14–15, 154–156, 159–164, 167–168, 171–172; in Trinidad 10, 80, 84–87
Renewable Energy and Energy Efficiency Agency (REEEA) 85–87

Renewable Energy Rider (RER) 144–145, 147
renewable energy targets (RETs) **84**, 87, 145, 191, 198, 200
renewable portfolio standards (RPSs) 14, 161–164, 172, 191, 198, 200
residential sector 36, 79, 126, 159, 166
resisting change 81; during energy transitions 29; and culture 52, 188–190; in the regime 32, 86, 173
role (s) of 'culture' 7, 17–18, 52, 58, 60–61, 188
rural 107–108, 195

sale 128, 134, 144, 173
science 6, 30–31, 40, 61, 85, 189
Small Island Developing State (SIDS) 10
small island (s) 4–6, 34, 36, 69, 81
social inequity 166–167, 173
Social Practice Theory (SPT) 53–54
socio-technical structures: and the energy cultures framework 58, 60–61; and grid-connected solar energy 166, 201; in the multi-level perspective 30, 32–33, 35, 40, 52, 55; narrative 127, 199; systems 29; regime resistance in Trinidad 81; *see also* energy transitions; landscape; regime
solar hot water 192; in Barbados 125, 127, 129–132, 134, **136**, 138–139, 141, 146; in Oʻahu 154
solar lighting 114, 136
solar power 36, 165, 168, 171–173, 194–198, 201
solar water heater (s) (SWH (s)) 8–9, 15, 17, 37, 39–40, 49, 57, 189–190, 192–194, 197, 200–201; in Barbados 12–13, 125–134, 136, 138–143, 145–147; in Oʻahu 153–154; in Trinidad 83–**85**, 87, 103–104, 112–114, 116–117; *see also* solar hot water
standard (s) 10, 69, 82; in policies 56, **85**, 146, 161–162, 170, 192–193, 199–200; *see also* certification; RPSs
subsidies 12, 56, 198; on energy in Trinidad 73–74, 76–77, 81–82, 87, 191; and grid-connected solar power 166; for innovations 33
sugarcane 69–70, 75, 86
Sustainability Cultures Framework 53

tariff (s) 35, 79, 106–107, 198; rates in Barbados 126; and grid-connected solar 164, 166–167, 169–170, 194–196, 198;

208 *Index*

tariff (s) *continued*
  rates in Oʻahu 157; rates in Trinidad 77; *see also* FiT
tax incentive (s)/credit (s) schemes: in Barbados 145, 147, 192; in Oʻahu 14, 161, 153–155, 167, 169, 172; tax credit (s) 194; tax incentive (s) 192; in Trinidad **84**, 87
the mainstream 28, 51; as an adjective 110, 134, 146; structure in energy transitions 33, 52, 58; theoretical definition 37–40, 54, 187
theory 15, 17, 75, 81, 108, 125, 154, 199; and energy transitions 30, 37, 40; and culture 53, 54, 60; *see also* Energy Cultures Framework (ECF); Multi-Level Perspective of Socio-technical Energy Transitions
timeframe (s) 34, 75, 86, 102, 159
transmission 55; infrastructure **97**, 110; and solar power 145, 156, 172; and utility companies 75–76, 79, 87, 98, 126; *see also* distribution network
Trinidad 8–9, 12–13, 15, 17, 35, 39, 167, 173, 188–189, 191–194, 197, 200–201; Barbados energy cultures comparison 125–126, 134, 136–137, 138, 142–146; case study background 9–10, 69–77, 81–84, 87; energy cultures and solar energy in 94–96, 98, 101, 103–104, 107, 109, 115–116; *see also* T&T
Trinidad and Tobago (T&T) 9–11, 13, 36, 78, 80, 98

Trinidad and Tobago Electricity Commission (T&TEC) authority: customers 11; and energy culture 95, **97–98**, 106, 111, 114–115; and renewable energy **84**, 85; role of 75–77, 79, 87, 94, 125–126
Trinidad Generation Unlimited (TGU) 76
tropical depressions 9, 11; *see also* hurricane (s)
Tropics 5, 8

United States of America (U.S.) 13, 155, 161
University of Hawaii (UofH) 159–160
University of the West Indies (UWI) 78, 80, 85, 126, 143
University of Trinidad and Tobago (UTT) 78, 80, 85
utility company 36, 161, 164, 166–167, 169, 171, 173, 189, 192, 195, 198; in Barbados 125, 127, 138, 145–146; in Oʻahu 162; in Trinidad 11, 94, **97–98**; *see also* Barbados Light and Power Company; Hawaiian Electric Company; Trinidad and Tobago Electricity Commission
utility-scale photovoltaics (PV) 145, 155–156

value added tax (VAT) **84**

wet season 9; *see also* rain
World War (s) (WW (s)) 71